# 1000 NATURE WORDS

## 兒童英漢圖解自然科學1000字

朱爾斯·波特 ◆ 著

新雅文化事業有限公司
www.sunya.com.hk

# 1000 NATURE WORDS
## 兒童英漢圖解自然科學 1000 字

作者：朱爾斯·波特（Jules Pottle）

翻譯：羅睿琪

責任編輯：張雲瑩

美術設計：黃觀山

出版：新雅文化事業有限公司

香港英皇道499號北角工業大廈18樓

電話：（852）2138 7998

傳真：（852）2597 4003

網址：http://www.sunya.com.hk

電郵：marketing@sunya.com.hk

發行：香港聯合書刊物流有限公司

香港荃灣德士古道220-248號荃灣工業中心16樓

電話：（852）2150 2100　傳真：（852）2407 3062

電郵：info@suplogistics.com.hk

版次：二〇二二年四月初版

ISBN: 978-962-08-7936-4

Original Title: *1000 WORDS NATURE*

Copyright © 2022 Dorling Kindersley Limited

A Penguin Random House Company

Traditional Chinese Edition © 2022 Sun Ya Publications (HK) Ltd.

18/F, North Point Industrial Building, 499 King's Road, Hong Kong

Published in Hong Kong, China

Printed in China

**For the curious**
www.dk.com

# 1000 NATURE WORDS

## 兒童英漢圖解自然科學1000字

朱爾斯．波特 ◆ 著

# 給爸爸媽媽的話

## 認識自然科學的重要性

在這個科技發達的世代裏，懂得大自然的美麗與奧妙，是維持我們身心健康的重要一環。如果你能花時間聽一聽、看一看身邊的事物，你便會發現千姿百態的自然奇觀，即使在城市的大街小巷裏，世界也正與生機勃勃的植物與動物共存，而地球上的每一處棲息地，以至每一種生物都各自有其重要之處。

年幼的孩子是天生的探險家，他們喜歡觸摸、嗅聞，讓自己沉醉在周遭的環境之中。本書將讓他們掌握與大自然有關的關鍵詞彙，用以分享他們的經驗，建立與野生生物、環境與地球相關的知識。本書亦將鼓勵孩子發問，引領他們看得更遠，學得更多。

本書中的「大象」與「長頸鹿」都是伴隨着我長大的熟悉詞彙，不過我的孫兒可能永遠無法在現實世界中看見這些動物。我們已面臨關乎地球未來的關鍵時刻，我們現今對大自然作出的抉擇，將影響我們的子女，甚至他們的子女，因此，能夠充分認識我們對地球的影響，以及我們能做什麼來保護地球，至關重要。本書正是一個很好的起始點，讓孩子開始學會欣賞大自然，並從小認知保育地球的重要性。

朱爾斯・波特 (Jules Pottle)
英國資深小學科學科顧問、教師、導師及作家

# Contents 目錄

## 安全提示

戶外休閒活動均具有潛在風險，幼兒參與本書中介紹的戶外活動時，家長需要從旁協助及指導。所有人均應承擔自身行動的責任，慎防發生意外，以獲得更安全及愉快的體驗。

# Our Planet 我們的星球

Earth orbits the Sun in space. Inside Earth is a hot core of liquid rock. Around the outside of Earth there are gases, called Earth's atmosphere. Earth formed a very long time ago. Its landscape and wildlife have changed over time.

地球在太空裏圍繞着太陽旋轉。地球的深層是熾熱的岩漿，而包圍在地球外面的是一些氣體，稱為大氣層。地球在很久很久以前形成，它的景觀和居住其中的野生生物隨時間過去而有所轉變。

**Earth**
地球

outer space
外太空

Is most of Earth's surface covered with land or with water?
地球表面大部分面積是給陸地、還是水覆蓋着的？

**Moon**
月球

land
陸地

## Some landscape features
部分地貌特徵

crater
火山口

volcano
火山

mountain
高山

valley
山谷

river
河流

lake
湖泊

geyser
間歇泉

spring
泉

rock
岩石

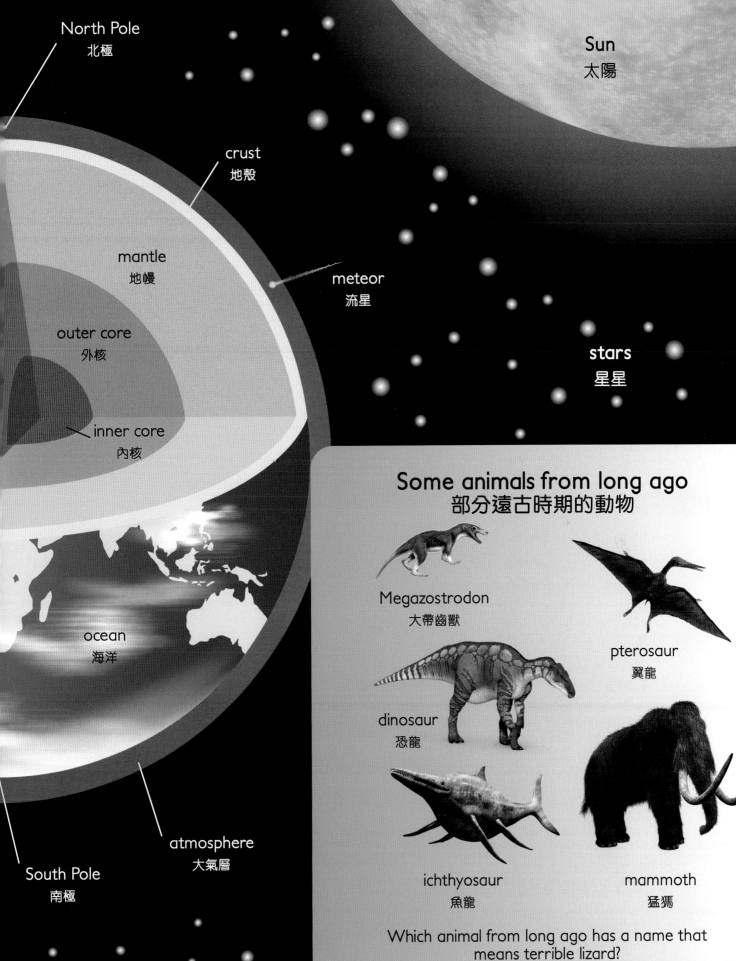

North Pole
北極

crust
地殼

mantle
地幔

outer core
外核

inner core
內核

ocean
海洋

atmosphere
大氣層

South Pole
南極

Sun
太陽

meteor
流星

stars
星星

## Some animals from long ago
## 部分遠古時期的動物

Megazostrodon
大帶齒獸

pterosaur
翼龍

dinosaur
恐龍

ichthyosaur
魚龍

mammoth
猛獁

Which animal from long ago has a name that
means terrible lizard?

哪一種遠古動物的英文名字意思是「可怕的蜥蜴」？

7

# Our world's resources 地球的資源

Our world is filled with useful materials. We must try not to waste them. We also need to reduce the pollution that some materials cause.

地球上充滿有用的物料，我們必須避免浪費它們，也需要減少因使用這些物料所造成的污染。

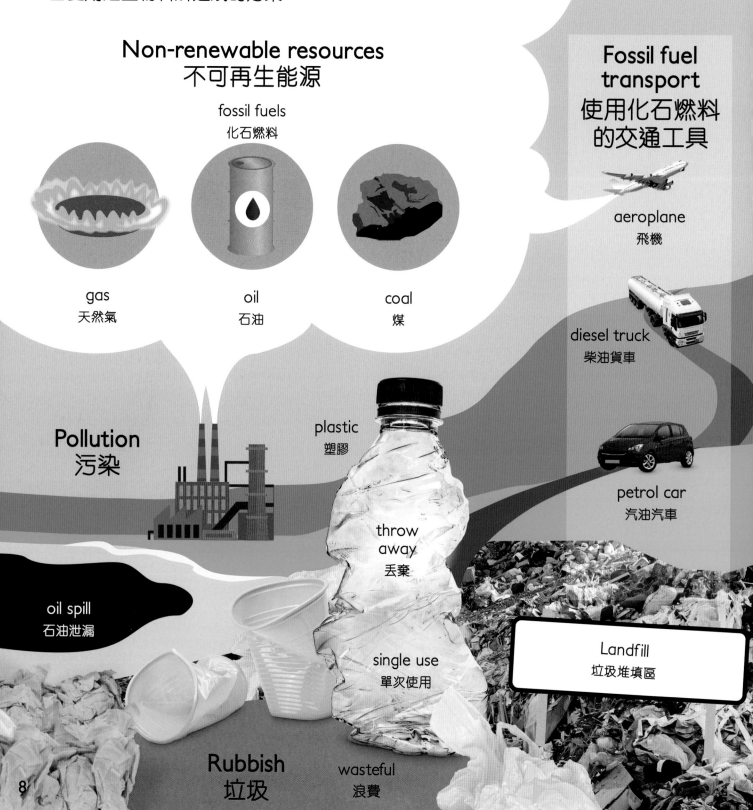

## Non-renewable resources
## 不可再生能源

fossil fuels
化石燃料

gas
天然氣

oil
石油

coal
煤

Fossil fuel transport
使用化石燃料的交通工具

aeroplane
飛機

diesel truck
柴油貨車

petrol car
汽油汽車

Pollution
污染

plastic
塑膠

throw away
丟棄

oil spill
石油泄漏

single use
單次使用

Landfill
垃圾堆填區

Rubbish
垃圾

wasteful
浪費

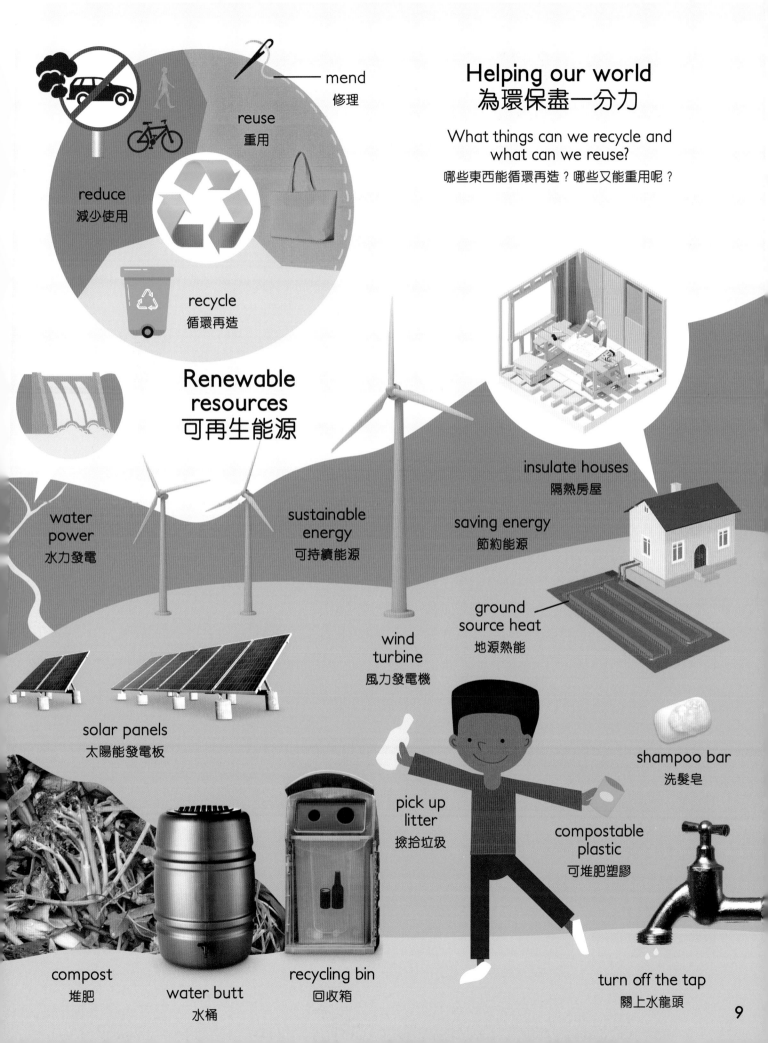

mend
修理

reuse
重用

reduce
減少使用

recycle
循環再造

## Helping our world
## 為環保盡一分力

What things can we recycle and
what can we reuse?
哪些東西能循環再造？哪些又能重用呢？

## Renewable
## resources
## 可再生能源

insulate houses
隔熱房屋

water
power
水力發電

sustainable
energy
可持續能源

saving energy
節約能源

ground
source heat
地源熱能

wind
turbine
風力發電機

solar panels
太陽能發電板

shampoo bar
洗髮皂

pick up
litter
撿拾垃圾

compostable
plastic
可堆肥塑膠

compost
堆肥

water butt
水桶

recycling bin
回收箱

turn off the tap
關上水龍頭

# Humans and nature 人與大自然

For thousands of years, we humans have made nature work for us. We should do this respectfully, taking care of our planet and the things that live on it.

數千年來，大自然一直在為我們人類效勞，在消耗大自然的物資時，我們應該心懷尊重，好好看顧我們的地球，並照顧生活在地球上的生物。

## On the farm 農場裏

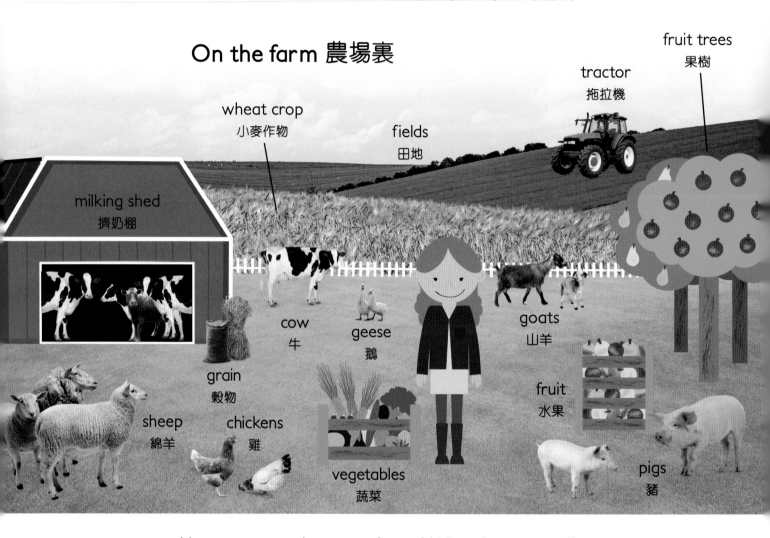

fruit trees
果樹

tractor
拖拉機

wheat crop
小麥作物

fields
田地

milking shed
擠奶棚

cow
牛

geese
鵝

goats
山羊

grain
穀物

fruit
水果

sheep
綿羊

chickens
雞

vegetables
蔬菜

pigs
豬

Have you ever taken care of a pet? What do pets need?
你有照顧過寵物嗎？寵物需要什麼呢？

## Pets 寵物

kitten
小貓

rabbit
兔子

dog
狗

goldfish
金魚

hamster
倉鼠

fish farm
養魚場

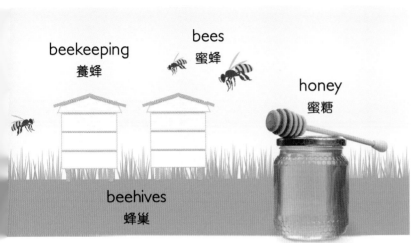

beekeeping
養蜂

bees
蜜蜂

honey
蜜糖

beehives
蜂巢

## Unusual farm animals
## 罕見的農場動物

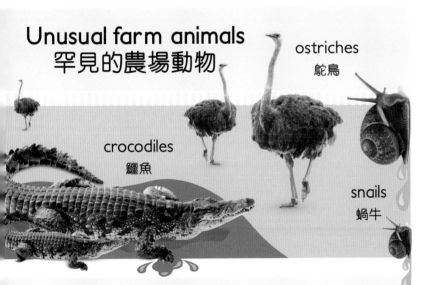

ostriches
鴕鳥

crocodiles
鱷魚

snails
蝸牛

## Working animals
## 役畜

sheepdog
牧羊犬

police horse
警馬

guide dog
導盲犬

## More types of farming
## 更多農業類別

flower farming
種植花卉

rice farming
種植稻米

vineyard
葡萄園

tea plantation
茶園

Farms around the world grow different crops. Do you know what's grown in a vineyard and used to make wine?
世界各地的農場栽種了不同的農作物。你知道葡萄園裏種植了什麼農作物用來釀酒嗎？

## Working with water
## 水利設施

dam
水壩

canal
運河

## Mining 採礦

limestone quarry
石灰岩採石場

coal mine
煤礦

11

# Nature activities
## 戶外活動

There are so many fun things we can do when we go out and enjoy nature.

在戶外郊遊時，有許多有趣的事情可以做呀。

nature spotting
觀察大自然

bark rubbing
樹皮拓印

nature trail
自然步道

binoculars
雙筒望遠鏡

building a den
建造秘密基地

tree hugging
擁抱樹木
touching
觸摸

splashing in puddles
跳水窪

pond dipping
在池塘邊撈魚

net
魚網

nature art
自然藝術

collecting
蒐集物品

fishing
釣魚

Forest school
森林學校

reading a map
閱讀地圖

We could start a nature collection. What things might we collect?
我們可以蒐集大自然的東西珍藏，你會蒐集什麼東西呢？

binning litter
將廢物扔進垃圾箱

paddling
嬉水

flying a kite
放風箏

building a
sandcastle
建造沙城堡

crabbing
捕蟹

sledging
滑雪橇

throwing
snowballs
拋雪球

making a
snow angel
製作雪天使圖形

snow
person
雪人

Let's choose some things we'd like to do when we go out.
齊來挑選一些我們在戶外時想要做的事情吧。

camping
露營

storytelling
說故事

stargazing
觀星

night walk
夜行

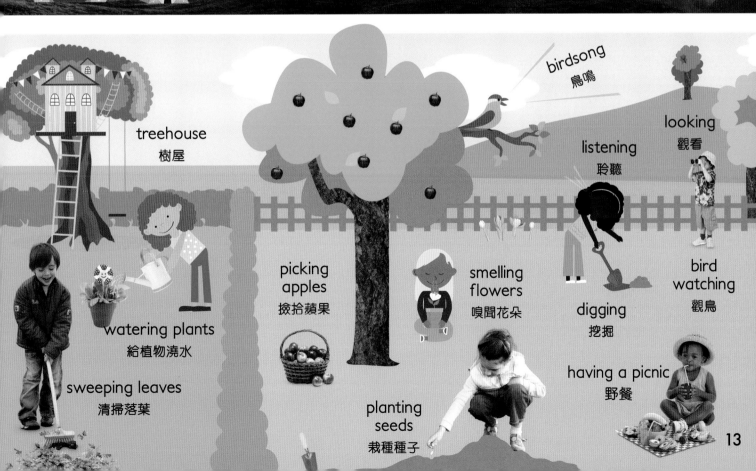

treehouse
樹屋

birdsong
鳥鳴

looking
觀看

listening
聆聽

picking
apples
撿拾蘋果

smelling
flowers
嗅聞花朵

bird
watching
觀鳥

digging
挖掘

watering plants
給植物澆水

sweeping leaves
清掃落葉

planting
seeds
栽種種子

having a picnic
野餐

13

# Weather 天氣

The weather is different in different places. Some countries have four seasons; others have two.

不同的地方有不同的天氣，有些地區擁有四個季節，有些地區則有兩個季節而已。

## Sunny
## 陽光普照

hot
炎熱

heatwave
熱浪

sun hat
太陽帽

sun cream
太陽油

blue sky
藍天

## Wet
## 潮濕

rainbow
彩虹

drizzle
毛毛雨

thunderstorm
雷暴

lightning 閃電

fog
霧

umbrella
傘子

rain
雨

puddles
水窪

## Cold
## 寒冷

hail 雹

snowstorm
暴風雪

snow
雪

icicles
冰柱

snowflakes
雪花

woolly hat
羊毛帽

ice
冰

frozen
結冰

frost
霜

14

## Cloudy 多雲

cirrus clouds
卷雲

overcast
密雲

stratus clouds
層雲

cumulus clouds
積雲

## Windy 大風

## The water cycle 水循環

Clouds form (made of water drops).
雲朵形成（由水點組成）。

Rain, snow, or hail falls.
雨、雪或雹降下。

Water vapour cools.
水蒸氣冷卻。

Water vapour forms.
水蒸氣形成。

Water (from rain, snow or hail) flows into rivers and seas.
水（來自雨、雪或雹）流進河流和海洋裏

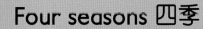

Liquid water changes into gas.
液態的水變成氣態。

## Four seasons 四季

spring 春

summer 夏

autumn 秋

winter 冬

## Two seasons 兩季

dry season
旱季

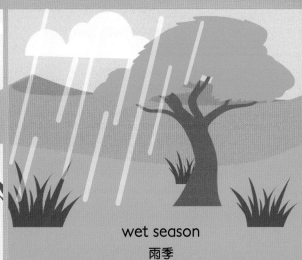

wet season
雨季

What type of weather do you like best?
你最喜歡哪種天氣？

15

# Extremes 極端環境

From hurricanes and floods to dry deserts and frozen lands, our planet has some extreme weather and some incredible locations.
從颶風與洪水，到乾旱的沙漠與冰封的土地，我們的星球擁有着極端天氣與不可思議的地方。

Spot a type of foggy weather that is caused by air pollution.
試指出一種由空氣污染導致的多霧天氣。

## Extreme weather 極端天氣

drought 乾旱  arid 乾燥

flood 洪水

tsunami 海嘯

cyclone 氣旋

hurricane 颶風

tornado 龍捲風

smog 煙霧

sandstorm 沙暴

extreme heat and wildfire 極端高温與山火

hailstorm 雹暴

blizzard 雪暴

ice storm 冰暴  freezing 凍結

# Incredible locations
## 不可思議的地方

### Dead Sea, Asia
### 亞洲 死海

lowest land area on Earth
地球上最低的陸上區域

### Mount Everest, Asia
### 亞洲 珠穆朗瑪峯

highest land area on Earth
地球上最高的陸上區域

### Furnace Creek, North America
### 北美洲 火爐溪

hottest recorded temperature
有紀錄以來最熱氣溫

### Vostok Station, Antarctica
### 南極洲 沃斯托克站

coldest recorded temperature
有紀錄以來最冷氣溫

### Atacama Desert, South America
### 南美洲 阿塔卡馬沙漠

driest place
最乾旱的地方

### Mawsynram, Asia
### 亞洲 毛辛拉姆

has highest rainfall
擁有最高降雨量

### Angel Falls, South America
### 南美洲 安赫爾瀑布

highest waterfall
最高的瀑布

### Grand Canyon, North America
### 北美洲 大峽谷

longest canyon
最長的峽谷

### Kilauea Volcano, Oceania
### 太平洋 基拉韋厄火山

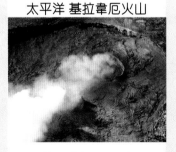

Earth's most active volcano
地球上最活躍的火山

### Great Barrier Reef, Pacific Ocean
### 太平洋 大堡礁

largest coral reef
最大的珊瑚礁

### Mariana Trench, Pacific Ocean
### 太平洋 馬里亞納海溝

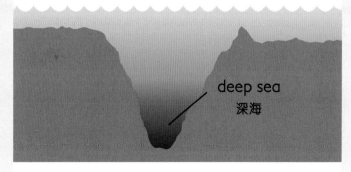

deep sea
深海

deepest part of the ocean 海洋最深的部分

Which of these amazing places would you like to explore?
你想到以上哪些令人驚歎的地方探索？

# Kingdoms of living things 生物王國

Different features help scientists put all living things into groups. Here are the five majoy groups, called kingdoms, and some of the smaller groups that are within each kingdom.

不同的特徵幫助科學家將所有生物分成不同類別，以下是五個主要類別，稱為界，而每一個界裏面又劃分成較小的類別。

## Plants 植物

### Flowering plants 開花植物

apple tree
蘋果樹

rose
玫瑰

grass
草

### Non-flowering plants 無花植物

conifer
針葉樹

fern
蕨類

mosses and liverworts
苔類及蘚類

## Fungi 真菌

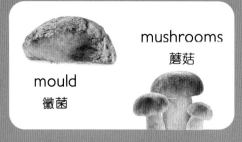

mushrooms
蘑菇

mould
黴菌

## Protists 原生生物

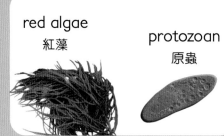

red algae
紅藻

protozoan
原蟲

## Monera 原核生物

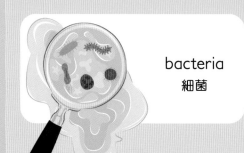

bacteria
細菌

## Molluscs 軟體動物

snail
蝸牛

oysters
蠔

## Crustaceans 甲殼類動物

woodlouse
潮蟲

crab
蟹

## Arachnids 蛛形綱動物

spider
蜘蛛

tick
壁蝨

## Cephalopods 頭足類動物

squid
魷魚

octopus
章魚

# Animals 動物

## Mammals
### 哺乳類動物

dolphin
海豚

kangaroo
袋鼠

human
人類

## Echinoderms
### 棘皮動物

starfish
海星

sea urchin
海膽

## Cnidarians
### 刺胞動物

jellyfish
水母

coral
珊瑚

## Birds
### 鳥類

robin
知更鳥

penguin
企鵝

## Reptiles
### 爬蟲類動物

snake
蛇

crocodile
鱷魚

## Amphibians
### 兩棲類動物

newt
蠑螈

frog
青蛙

## Poriferans
### 多孔動物

sponge
海綿

Let's think of some
more mammals and
some more reptiles.

我們想一想還有哪些哺乳類
動物和爬蟲類動物吧。

## Fish
### 魚類

clownfish
小丑魚

shark
鯊魚

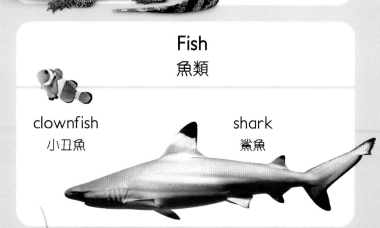

## Insects
### 昆蟲

ant
螞蟻

butterfly
蝴蝶

## Myriapods
### 多足動物

millipede
馬陸

centipede
蜈蚣（百足）

## Annelids
### 環節動物

leech
水蛭

earthworm
蚯蚓

19

# All sorts of plants 各種各樣的植物

There are many different plants, but they all do something amazing – they use water, air, and sunlight to make their own food. This process is called photosynthesis.

世界上有許多不同的植物，不過它們全都會做一件令人嘖嘖稱奇的事情，它們會利用水、空氣和陽光來製造自己的食物，這個過程稱為光合作用。

Herbs can be used to flavor food. What herbs have you tried?
香草可以用來為食物調味，你嘗過哪些香草呢？

## Shrubs 灌木

hydrangea
繡球花

boxwood 黃楊木

## Herbs 香草

basil
羅勒

thyme 百里香

Guess where baobab trees store water.
猜猜看，猴麵包樹會將水儲存在哪裏？

## Flowers 花卉

hibiscus
木槿（大紅花）

orchid
蘭花

scent
香味

petal
花瓣

bud
花蕾

leaf
葉子

stalk
莖

thorn
刺

daffodils
水仙花

tulips
鬱金香

rose
玫瑰

## Weird plants
## 怪異的植物

venus
flytrap
捕蠅草

air plant
空氣草

catches flies!
它會捕捉蒼蠅！

doesn't need soil
它不需要生長在土壤中。

baobab tree
猴麵包樹

has a wide trunk
它擁有寬闊的樹幹。

## Climbers 攀緣植物

ivy
常春藤

runner bean
紅花菜豆

## Cacti 仙人掌

golden barrel
金琥仙人掌

bunny ear
兔耳仙人掌

## Trees 喬木

broadleaved
闊葉樹

conifers
針葉樹

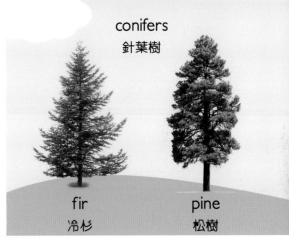

maple
楓樹

weeping willow
垂柳

palm
棕櫚樹

fir
冷杉

pine
松樹

## Fruit and vegetables we eat
## 我們食用的水果與蔬菜

orange
橙

apple
蘋果

kiwi fruit
奇異果

carrots
胡蘿蔔

cabbage
椰菜

broccoli
西蘭花

## Photosynthesis
## 光合作用

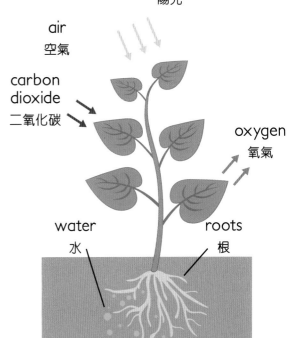

sunlight
陽光

air
空氣

carbon
dioxide
二氧化碳

oxygen
氧氣

water
水

roots
根

## Water plants 水生植物

seaweed
海藻

water lily
睡蓮

21

# A closer look at trees
## 喬木放大鏡

A tree is a tall plant with a thick stem called a trunk. Like all plants, trees improve air quality by absorbing carbon dioxide and giving out oxygen.

喬木是高大的植物，擁有粗壯的莖，稱為樹幹。就像所有植物一樣，喬木能藉由吸收二氧化碳，並釋出氧氣，從而改善空氣質素。

sun
太陽

oxygen
氧氣

pine needles
松針

pine cone
松果

twigs
幼枝

roots
根

## Tree leaves
## 樹葉

lobed
淺裂的

pointy
尖的

wavy edges
波浪形邊緣的

prickly
多刺的

needles
針狀葉的

### Parts of a leaf
### 葉子的各部分

tip
葉尖

midrib
主葉脈

vein
葉脈

margin
(edge)
葉緣

stem
葉莖

22

**evergreen** 常綠樹
keeps leaves all year
全年都長有樹葉。

pine tree
松樹

carbon
dioxide
二氧化碳

## Tree flower 喬木的花朵

cherry blossom
櫻花

hazel catkins
榛樹的菜荑花序

apricot buds
杏樹的花蕾

## Tree fruit 喬木的果實

apple
蘋果

plum
李子

lemon
檸檬

oak tree
橡樹

wood
木材

tree rings
年輪

oak leaf
橡樹葉

tree knot
樹結

branch
樹枝

bark
樹皮

trunk
樹幹

sticks
枝條

acorns
橡實

The seed of an oak tree is inside an acorn.
Where would you find the seeds of a pine tree?
橡樹的種子藏在橡實裏，你能在哪裏找到松樹的種子呢？

**deciduous** 落葉樹
loses leaves in autumn
會在秋天落葉。

23

# Plant and fungus life cycles

## 植物與真菌的生命周期

Most plants grow from seeds, bulbs, or tubers.
Fungi grow from tiny spores.

大部分植物都是從種子、鱗莖或塊莖中生長
出來的；真菌則從微細的孢子中生長出來。

pollinator
授粉者

flower
花朵

seeds
種子

germination
萌芽

Sunflower (plant)
向日葵（植物）

shoot
嫩芽

plant
植物

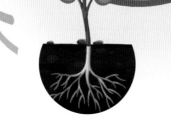

seedling
幼苗

Have you planted any seeds? How did you take care of them? Think of
some things you did to help them grow.

你試過種植種子嗎？你是如何照顧它們的？想一想你曾做過哪些事情來幫助種子生長吧。

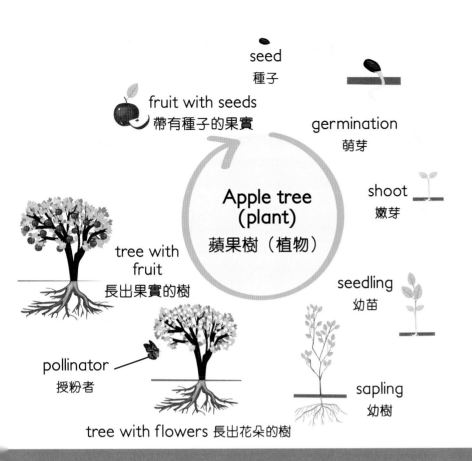

## Apple tree (plant)
蘋果樹（植物）

seed
種子

fruit with seeds
帶有種子的果實

germination
萌芽

shoot
嫩芽

seedling
幼苗

sapling
幼樹

tree with flowers 長出花朵的樹

pollinator
授粉者

tree with fruit
長出果實的樹

## Pollination
授粉

pollen grains
花粉粒

pollen
花粉

pollinator
授粉者

stigma
柱頭

ovary
子房

seed
種子

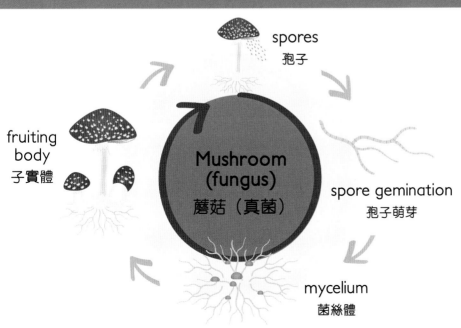

## Mushroom (fungus)
蘑菇（真菌）

spores
孢子

spore gemination
孢子萌芽

mycelium
菌絲體

fruiting body
子實體

## More seeds
更多不同的種子

seeds in woody scales
藏在木質鱗片中的種子

pine cone
松果

sycamore seeds
美國梧桐種子

avocado stone
牛油果核

seeds (peas)
種子（豌豆）

pea pods
豌豆莢

## Bulbs and tubers
鱗莖及塊莖

tulip bulbs
鬱金香鱗莖

potato tuber
馬鈴薯塊莖

# Animal life cycles
## 動物的生命周期

Cats have kittens, and the kittens grow into cats. Then those cats have kittens. Let's learn more about animal life cycles.

貓會生幼貓，而幼貓會長大成貓，然後那些貓也會生幼貓。我們來認識一下動物的生命周期吧。

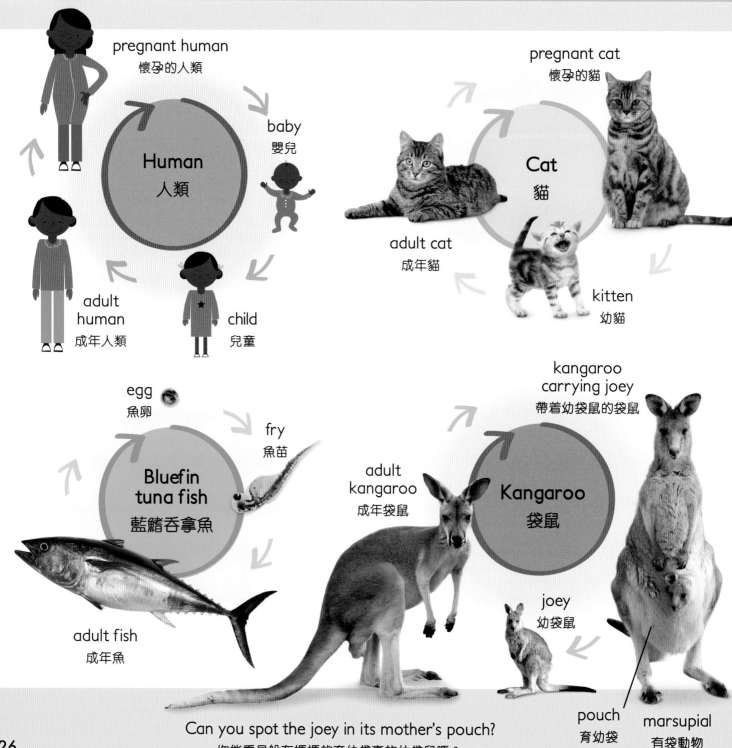

pregnant human
懷孕的人類

baby
嬰兒

**Human**
人類

adult human
成年人類

child
兒童

pregnant cat
懷孕的貓

**Cat**
貓

adult cat
成年貓

kitten
幼貓

egg
魚卵

fry
魚苗

**Bluefin tuna fish**
藍鰭吞拿魚

adult fish
成年魚

kangaroo carrying joey
帶着幼袋鼠的袋鼠

adult kangaroo
成年袋鼠

**Kangaroo**
袋鼠

joey
幼袋鼠

pouch
育幼袋

marsupial
有袋動物

Can you spot the joey in its mother's pouch?
你能看見躲在媽媽的育幼袋裏的幼袋鼠嗎？

26

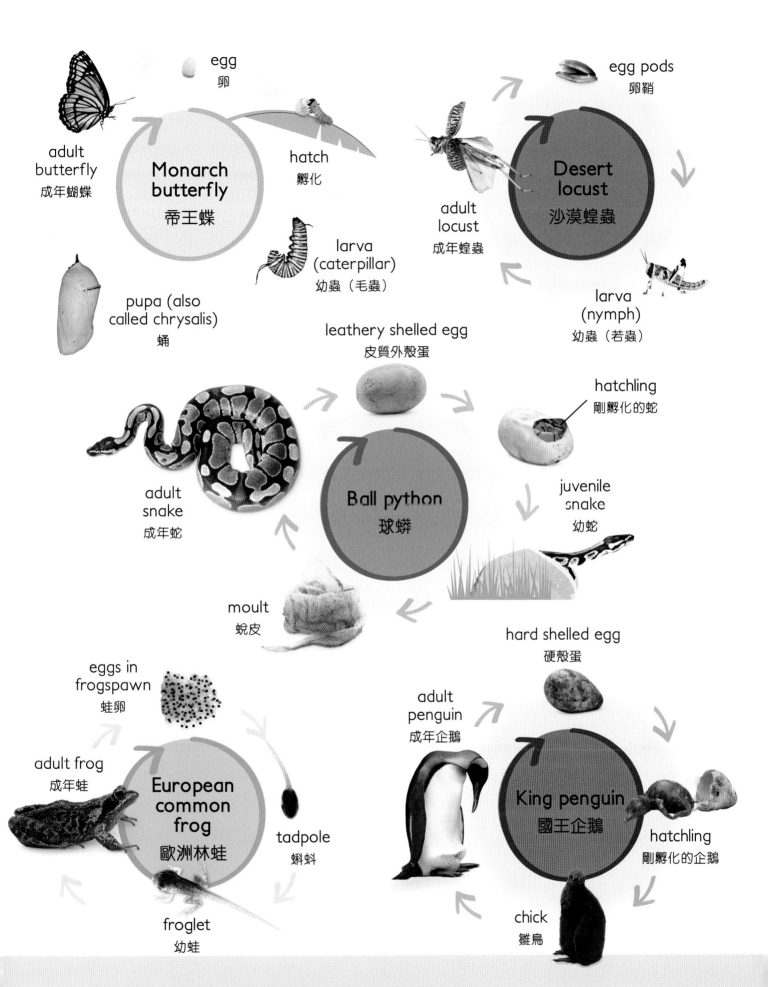

egg
卵

hatch
孵化

Monarch
butterfly
帝王蝶

adult
butterfly
成年蝴蝶

larva
(caterpillar)
幼蟲（毛蟲）

pupa (also
called chrysalis)
蛹

egg pods
卵鞘

Desert
locust
沙漠蝗蟲

adult
locust
成年蝗蟲

larva
(nymph)
幼蟲（若蟲）

leathery shelled egg
皮質外殼蛋

hatchling
剛孵化的蛇

adult
snake
成年蛇

Ball python
球蟒

juvenile
snake
幼蛇

moult
蛻皮

hard shelled egg
硬殼蛋

eggs in
frogspawn
蛙卵

adult
penguin
成年企鵝

adult frog
成年蛙

European
common
frog
歐洲林蛙

King penguin
國王企鵝

tadpole
蝌蚪

hatchling
剛孵化的企鵝

froglet
幼蛙

chick
雛鳥

Do you know where frogs lay their frogspawn?
你知道青蛙會在哪裏產卵嗎？

27

# Animal families 動物家族

Can you name the female, male, and baby in each of these animal families? Some have special names and others don't.

你知道這些動物家族中的雌性、雄性和小寶寶叫什麼名字嗎？牠們有些擁有特別的名稱，有些則沒有。

bull
公牛

mare
母馬

stallion
公馬

calf
牛犢

cow
母牛

foal
幼駒

boar
公豬

cockerel
公雞

ewe
母羊

ram
公羊

lambs
羔羊

sow
母豬

piglet
小豬

hen
母雞

chicks
小雞

bitch
母狗

male ladybird
雄性瓢蟲

female ladybird
雌性瓢蟲

tom
雄貓

cat
雌貓

larvae
幼蟲

kitten
小貓

puppy
小狗

dog
公狗

drake
公鴨

duck
母鴨

duckling
雛鴨

male
spider
雄性蜘蛛

female spider
雌性蜘蛛

spiderlings
幼蜘蛛

buck or boomer
公袋鼠

doe or
flyer
母袋鼠

joey
幼袋鼠

doe
母鹿

stag
公鹿

boar
公熊貓

fawn
幼鹿

sow
母熊貓

cub
幼熊貓

man
男人

woman
女人

male turtle
雄性龜

female
turtle
雌性龜

baby
嬰兒

hatchling 剛孵化的龜

## Animal groups
動物族羣

What names for groups of animals do you know?
你認識哪些動物族羣的名稱？

pride of lions
獅羣

pod of dolphins
海豚羣

flock of sheep
綿羊羣

colony of ants
蟻羣

# Heads, bodies, and feet 頭、身體和腳

Think of an animal. Imagine its head, body, and feet. Are they the same as yours?
試回想任何一隻動物的頭、身體和腳，牠的這些部分和你的長得一模一樣嗎？

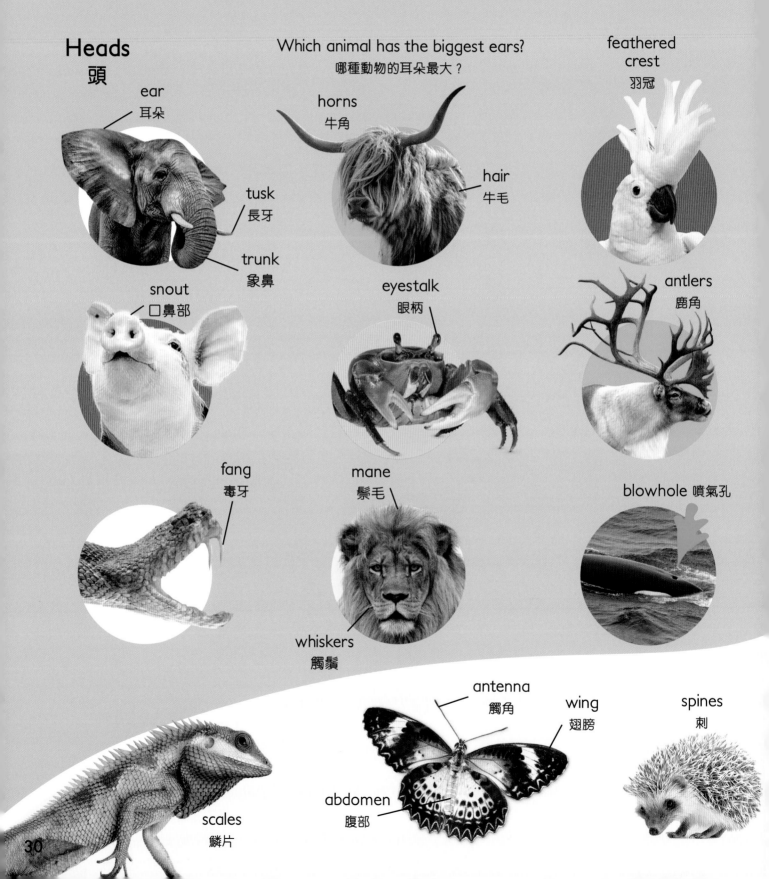

## Heads
頭

ear
耳朵

tusk
長牙

trunk
象鼻

snout
口鼻部

fang
毒牙

Which animal has the biggest ears?
哪種動物的耳朵最大？

horns
牛角

hair
牛毛

eyestalk
眼柄

mane
鬃毛

whiskers
觸鬚

feathered crest
羽冠

antlers
鹿角

blowhole 噴氣孔

antenna
觸角

wing
翅膀

spines
刺

abdomen
腹部

scales
鱗片

# Feet (and hands!) 腳（還有手！）

tube feet
管足

flipper
鰭肢

paws
掌

talons
鷹爪

hooves
蹄

sticky pads
黏性腳墊

prolegs
原足

webbed feet
蹼足

claws
爪子

fin
鰭

fingers
手指

toes
腳趾

Why do waterbirds have webbed feet?
為什麼水鳥長有蹼足？

skin
皮膚

## Bodies
身體

fur
毛皮

four-legged
四肢

shell
殼

tail
尾巴

bristles
豬鬃

feathers
羽毛

exoskeleton (outside skeleton)
外骨骼

endoskeleton (inside skeleton)
內骨骼

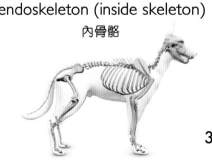

# Feeding time 用餐時間

What do animals feed on and how do they eat?
Let's take a look.

動物會進食什麼食物？牠們又會如何進食？
我們一起來看看吧。

**omnivore (plant and meat eater)**
雜食動物（會吃植物和肉的動物）

bears
熊

**herbivore (plant eater)**
草食動物（主要吃植物的動物）

**carnivore (meat eater)**
肉食動物（主要吃肉的動物）

predator
捕食者

cheetah
獵豹

vulture
禿鷲

graze
吃草

antelope
羚羊

**scavenger
(eater of leftovers)**
食腐動物
（主要吃殘餘食物的動物）

grass
草

prey
獵物

## Food chain
食物鏈

 flow of nutrients 營養轉移

## All about eating
飲食面面觀

mouth
嘴巴

**bite and chew**
咬嚙與咀嚼

Beavers eat leaves, twigs, and bark. They cut wood to make dams.
河狸會吃葉子、小枝條和樹皮，牠們會咬斷樹木來建造堤壩。

**gnaw**
啃咬

proboscis
(straw-like mouthpart)
吻管（一種像飲管的口器）

trap-jaw ants eat other insects.
鋸針蟻會吃其他昆蟲。

nectar
花蜜

mandibles
(mouthparts)
大顎（口器）

**suck**
吸食

**lap**
舔食

**ready to snap**
預備猛力一咬

# Animals with specialised diets
## 有特殊飲食習慣的動物

dung beetle
糞金龜

dung
糞便

panda
熊貓

bamboo
竹子

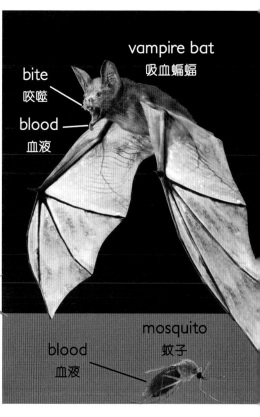

vampire bat
吸血蝙蝠

bite
咬嚙

blood
血液

mosquito
蚊子

blood
血液

Can you name another baby animal that feeds on its mother's milk?
你能說出另一種會喝媽媽乳汁的動物寶寶嗎？

beak
喙

crack!
咔啦！

nut
堅果

crush
壓碎

milk
乳汁

suckle
吮乳

whale baleen
(filtering mouthparts)
鯨鬚（負責過濾的口器）

krill
磷蝦

filter feeder (strains food)
濾食動物（會過濾出食物的動物）

Many spiders liquefy the insides of insects, then they suck up the liquid.
許多蜘蛛會將昆蟲的身體內部液化，然後將液體吸食掉。

liquefy
液化

# Communicating 溝通交流

Zoologists can't talk to animals, but they can understand some of their messages. Can you?

動物學家無法與動物交談，不過他們能明白動物的部分信息。你做得到嗎？

## Communicating over long distances
長距離通訊

low-pitched sound
低沉的聲音

## Marking territory
標記地盤

birdsong
鳥鳴

scent marking
氣味標記

alarm call
警示呼叫

hair on end
豎起毛髮

posture
姿勢

rearing up
用後腿直立

## Communicating danger
通報危險

## Showing fear or strength
展示恐懼或力量

growling
低吼

warning of attack
襲擊警告

howling
嚎叫

thumping
用力跺腳

## Communicating "I'm here"
通報「我在這裏」

## showing friendship
表示友好

grooming
梳理毛髮

cheep cheep
啾啾

call
呼叫

## Showing where food is
## 顯示食物的位置

Ants leave a scent trail.
螞蟻會留下氣味的痕跡。

Bees waggle dance (to tell other bees where there are flowers).
蜜蜂會來回舞動（以告訴其他蜜蜂花朵在哪裏）。

## Communicating with humans
## 與人類溝通

Koko the gorilla
大猩猩可可

sign language
手語

## Communicating "let's play!"
## 表達「我們來玩吧！」

A male bird of paradise puts on a display.
雄性天堂鳥正展開羽毛炫耀求偶。

A male cricket chirps.
雄性蟋蟀發出唧唧的叫聲。

play bow
玩躬

## Attracting a mate
## 吸引伴侶

Many male birds sing to impress a mate. Some female birds sing, too.
許多雄性鳥類會唱歌來取悅伴侶，有些雌性鳥類也會唱歌呢。

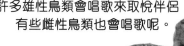

A male pufferfish creates a nest ready for a female's eggs.
雄性河豚造出巢穴，預備給雌性產卵。

Which animal wags its tail to show different emotions?
哪一種動物會搖擺尾巴來顯示不同的情緒？

35

## wild
### 野生的

wolves
狼

## tame
### 馴化的

cat
貓

## domesticated
### 畜養的

sheep
綿羊

# Animal comparison
## 動物大對碰

Big or small, wild or tame, venomous or harmless – just look at all these animal differences!

是大或小，野生或馴化，有毒或無害？快來看看動物的各種不同之處吧！

**fastest on land**
陸上最快的

cheetah
獵豹

**two of the slowest**
其中兩種最慢的

snail
蝸牛

sea anemone
海葵

What is the biggest animal that walks on land?
(Clue: it has a long trunk.)
在陸地上行走的最大動物是什麼？（提示：牠有一根長長的鼻子。）

blue whale
藍鯨

**heavy**
沉重

**light**
輕盈

## biggest
### 最大的

**smallest living things**
最小的生物

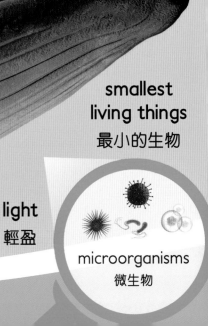

microorganisms
微生物

**solitary**
獨自活動的
snow leopard
雪豹

**social**
羣居的
dogs
狗

**diurnal (active in the daytime)**
晝行性的（活躍於日間）
butterfly
蝴蝶

**venomous**
有毒的
viper
毒蛇

**harmless**
無害的
slowworm
蛇蜥

**eye-catching**
惹人注目的
bird of paradise
天堂鳥

**camouflaged**
有保護色的
octopus
章魚

**nocturnal (active at night)**
夜行性的（活躍於晚間）
moth
蛾

**tall**
高的
giraffe
長頸鹿

**short**
矮的
mouse
老鼠

**plain**
素色的
horse
馬

**patterned**
有圖案的
zebra
斑馬

tiger
老虎

rabbit
兔子

fox
狐狸

**omnivore**
雜食動物

**carnivore**
肉食動物

**herbivore**
草食動物

## Poo 糞便

Poo comes in many shapes and sizes! Which animal has cube-shaped poo?

糞便有許多不同的形狀和大小！哪一種動物會排出立方體狀的糞便？

**rabbit droppings**
兔子糞便

**fox poo**
狐狸糞便

**bird dropping**
鳥糞

**lizard poo**
蜥蜴糞便

**wombat poo**
袋熊糞便

# Animal clues
## 動物線索

Learn to recognise the poo of different animals. Look out for rotting plants, too - these are being eaten by animals and microbes. You might also find evidence of things that lived long ago.

要找尋動物的線索，要從辨別不同動物的糞便學起，繼而留意腐爛的植物，因為這些植物曾給動物和微生物進食過，最後，從其他生物存在的證據也能找到更多動物線索。

## Rotting plants
### 腐爛的植物

What creatures are eating the log?

有哪些生物正在進食這棵倒下的樹木？

**microbes**
微生物

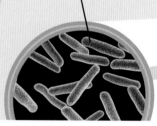

**animal tracks**
動物足跡

**bones**
骨頭

## More animal clues
### 更多動物線索

**shark teeth**
鯊魚牙齒

**shells**
貝殼

**skeleton**
骨骼

owl pellet
貓頭鷹糞便

otter spraint
水獺糞便

bat guano
蝙蝠糞便

Seeds can be transported in animal poo.
種子能夠經由動物糞便被傳播到其他地方。

wormcast
蚯蚓糞便

insect frass
昆蟲排泄物

bear scat
熊糞

elephant dung
大象糞便

natural compost
天然堆肥

dead
死亡

carbon dioxide
二氧化碳

log
原木

mushrooms
蘑菇

fungi
真菌

new growth
新生樹木

nutrients
營養素

leaf litter
落下的枯葉

natural recycling
自然循環

stag beetle larvae
鍬形蟲幼蟲

worm
蚯蚓

digest
消化

stag beetle
鍬形蟲

nuts buried by squirrels
松鼠埋藏的堅果

# Clues from long ago
# 很久以前的線索

fossil skull
頭骨化石

ammonite (fossil mollusc)
菊石（軟體動物化石）

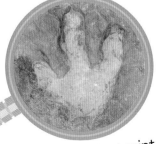

dinosaur footprint in fossil rock
在化石裏的恐龍足印

amber (fossil tree resin)
琥珀（樹脂化石）

coprolite (fossil poo)
糞化石（糞便化石）

# Gardens and parks 花園與公園

There are lots of wonderful things to see and do in a garden or park.

在花園或公園裏，可以看見許多奇妙的事物，還有許多有趣的事情可做。

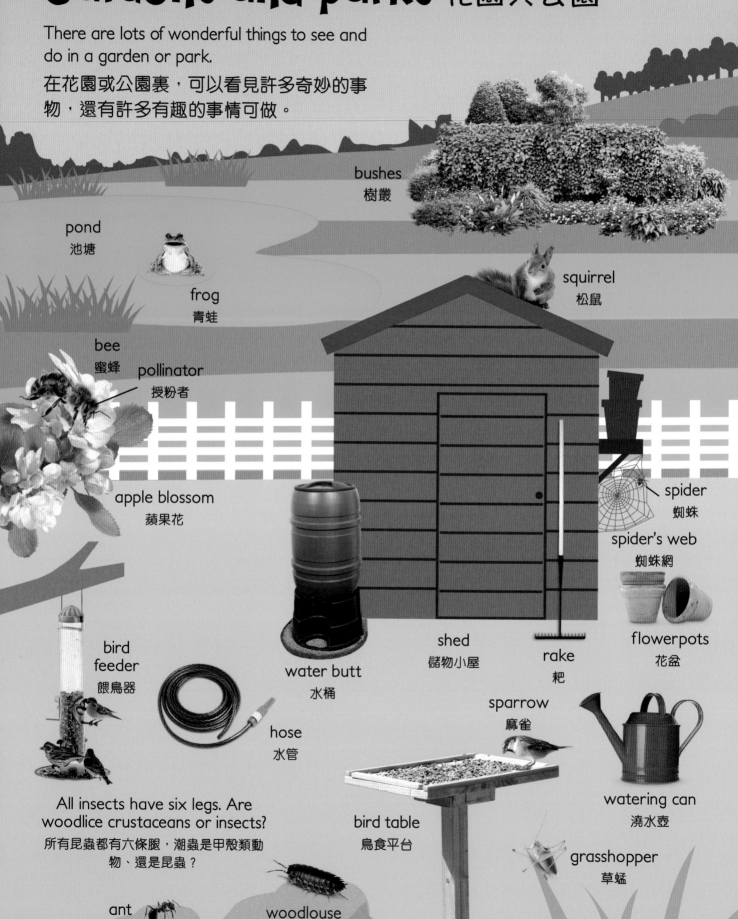

bushes
樹叢

pond
池塘

frog
青蛙

squirrel
松鼠

bee
蜜蜂

pollinator
授粉者

apple blossom
蘋果花

spider
蜘蛛

spider's web
蜘蛛網

flowerpots
花盆

bird feeder
餵鳥器

water butt
水桶

shed
儲物小屋

rake
耙

hose
水管

sparrow
麻雀

watering can
澆水壺

All insects have six legs. Are woodlice crustaceans or insects?

所有昆蟲都有六條腿，潮蟲是甲殼類動物，還是昆蟲？

bird table
鳥食平台

grasshopper
草蜢

ant
螞蟻

woodlouse
潮蟲

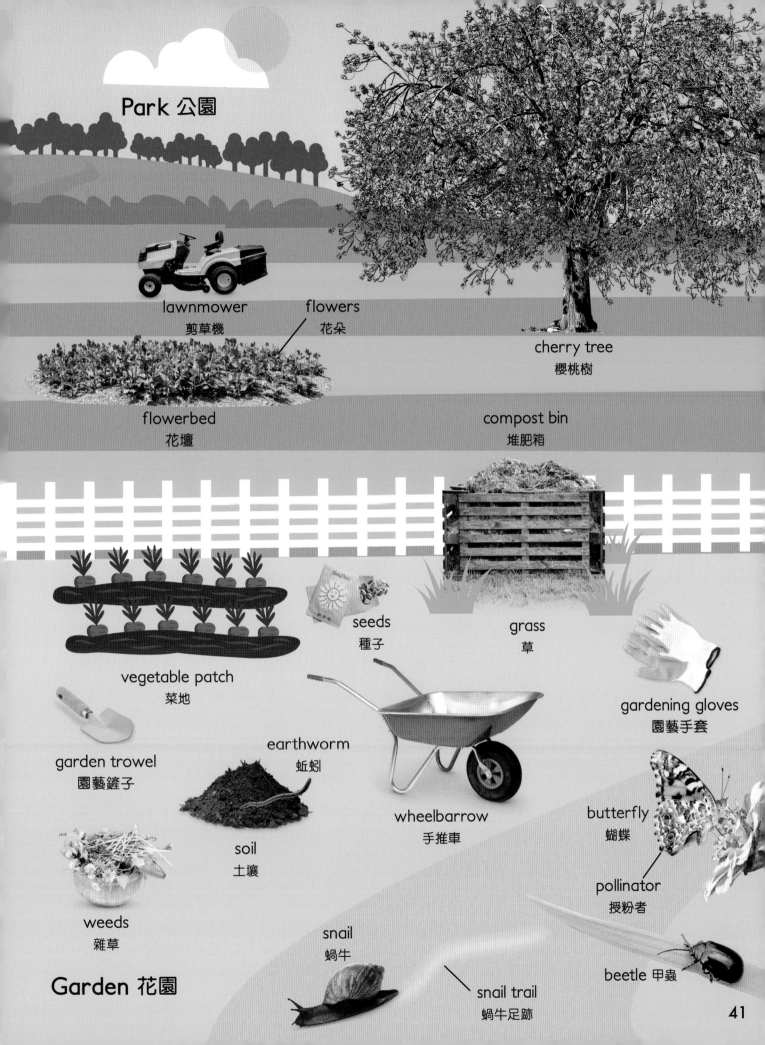

Park 公園

lawnmower
剪草機

flowers
花朵

cherry tree
櫻桃樹

flowerbed
花壇

compost bin
堆肥箱

seeds
種子

grass
草

gardening gloves
園藝手套

vegetable patch
菜地

garden trowel
園藝鏟子

earthworm
蚯蚓

wheelbarrow
手推車

butterfly
蝴蝶

soil
土壤

pollinator
授粉者

weeds
雜草

snail
蝸牛

beetle 甲蟲

Garden 花園

snail trail
蝸牛足跡

# Fields and meadows 田野與草原

Some fields are planted with crops. Others are full of wildflowers. Grassland can be green and grassy or dusty and dry.

有些田野栽種了農作物，也有些布滿了野花；草原可以是綠草如茵，又或是塵土飛揚及乾燥。

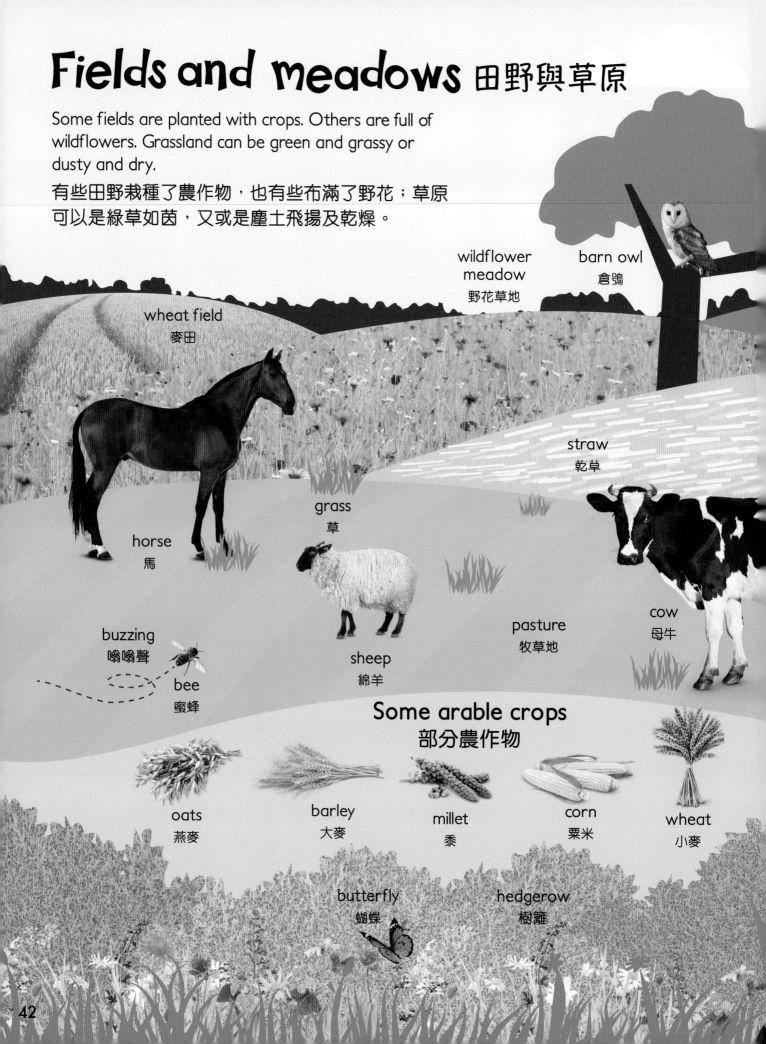

wildflower meadow
野花草地

barn owl
倉鴞

wheat field
麥田

straw
乾草

grass
草

horse
馬

buzzing
嗡嗡聲

bee
蜜蜂

sheep
綿羊

pasture
牧草地

cow
母牛

## Some arable crops
## 部分農作物

oats
燕麥

barley
大麥

millet
黍

corn
粟米

wheat
小麥

butterfly
蝴蝶

hedgerow
樹籬

Can you think of a food you like
that contains an arable crop?
你能想出一種你喜歡的食物，這食物是由
農作物製造而成的嗎？

osprey
魚鷹

baler
打包機

tractor
拖拉機

combine
harvester
聯合收割機

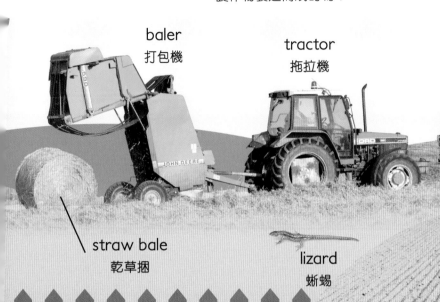

straw bale
乾草捆

lizard
蜥蜴

plough
犁

fence
圍欄

grey partridge
灰山鶉

field mouse
田鼠

grasshopper
草蜢

mole hole
鼴鼠洞

mole
鼴鼠

grass snake
水游蛇

clover
三葉草

mole
burrow
鼴鼠穴

rabbit burrow
兔穴

wild rabbit
野兔

# Woodlands 林地

Animals find plenty to eat in woodlands, and protection from the worst of the weather.

動物在林地裏能找到大量食物，還能找到躲避惡劣天氣的藏身之所。

horse chestnut leaf
七葉樹葉

conker shell
七葉樹果硬殼

conker
七葉樹果

chirping
啾啾

red squirrel
紅松鼠

chiffchaff
嘰喳柳鶯

hooting
嗚嗚

tawny owl
灰林鴞

bird's nest
鳥巢

undergrowth
林下植物

bluebells
風鈴草

horse chestnut tree
歐洲七葉樹

fawn
幼鹿

rustling
沙沙聲

ferns
蕨類

dragonfly
蜻蜓

pond
池塘

fox den
狐狸窩

fox
狐狸

pond snail
椎實螺

pond skater
水黽

rhinoceros beetle
兜蟲

## European forest
歐洲森林

Can you spot a baby deer?
你能找出鹿寶寶嗎？

44

# American redwood forest
## 美洲紅杉林

giant
redwood
tree
巨型紅杉樹

lichen
地衣

giant redwood cone
巨型紅杉樹毬果

giant redwood
leaves
巨型紅杉樹葉

bobcat
山貓

knocking
敲擊聲

woodpecker
啄木鳥

black bear
黑熊

toadstool 毒蕈

chipmunk
花栗鼠

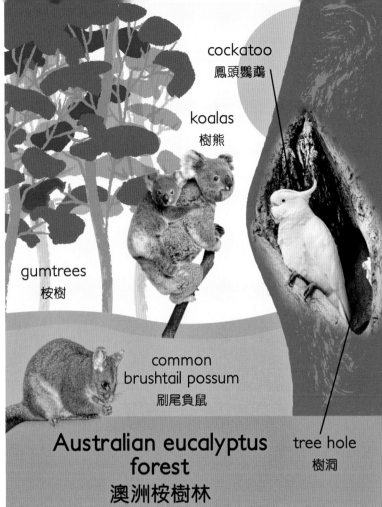

cockatoo
鳳頭鸚鵡

koalas
樹熊

gumtrees
桉樹

common
brushtail possum
刷尾負鼠

tree hole
樹洞

# Australian eucalyptus
## forest
## 澳洲桉樹林

bamboo
竹

plum
blossom
梅花

panda
熊貓

# Chinese bamboo forest
## 中國竹林

45

# Rivers 河流

Animals live in rivers, on riverbanks, and on the flat areas next to rivers, called floodplains.

動物會生活在河流中、河堤上、還有河邊平坦的地區裏，這些地方稱為沖積平原。

weeping willow tree
垂柳

burrow
獸穴

riverbank
河堤

muskrat
麝鼠

fishing
釣魚

Canada goose
加拿大雁

water vole
水䶄

osprey
魚鷹

fresh water
淡水

salmon
鮭魚

moorhen
紅冠水雞

carp
鯉魚

reeds
蘆葦

mayfly
蜉蝣

pondweed
水草

stickleback
刺魚

crayfish
螯蝦

pike
梭子魚

rapids
急流

current
水流

canoe
獨木舟

wild swimming
野外游泳

swan
天鵝

Spot the creature in the water that has an exoskeleton (a hard outer skeleton).
試在水中找出擁有外骨骼
（一種堅硬的外在骨骼）的生物。

duck
鴨子

heron
蒼鷺

dragonfly
蜻蜓

otter
水獺

beaver
河狸

marsh
沼澤

## Where does a river meet the sea?
河流會在哪裏和大海會合？

sea
大海

estuary
河口

river
河流

weir
堰

floodplain
沖積平原

waterfall
瀑布

47

# Oceans and coasts 海洋與海岸

Most of Earth's surface is covered with water. That is why it looks blue from space.
地球的表面大部分都被水覆蓋着，那就是為什麼從太空看下來地球是藍色的。

wind farm
風力發電場

sand
沙子

bay
海灣

surf
滑浪

swim
游泳

dive
潛水

wave
海浪

holiday
度假

## Warm, shallow sea
## 溫暖的淺海

octopus
章魚

Which sea animal has five long arms?
哪一種海洋動物擁有五根長長的手臂？

swordfish
劍魚

sea sponge
海綿

seahorse
海馬

clownfish
小丑魚

dolphin
海豚

seaweed
海藻

shells
貝殼

blue crab
藍蟹

coral
珊瑚

mussel
貽貝（青口）

pebbles
卵石

sea
anemone
海葵

brittle star
蛇尾

spiny lobster
龍蝦

48

What can you see that produces electricity?
你看見有什麼東西能產生電力？

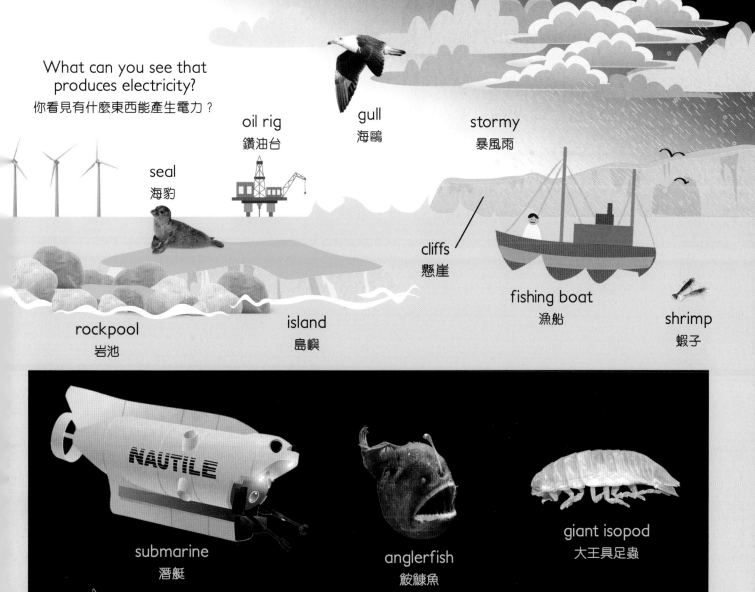

oil rig
鑽油台

gull
海鷗

stormy
暴風雨

seal
海豹

cliffs
懸崖

fishing boat
漁船

shrimp
蝦子

rockpool
岩池

island
島嶼

submarine
潛艇

anglerfish
鮟鱇魚

giant isopod
大王具足蟲

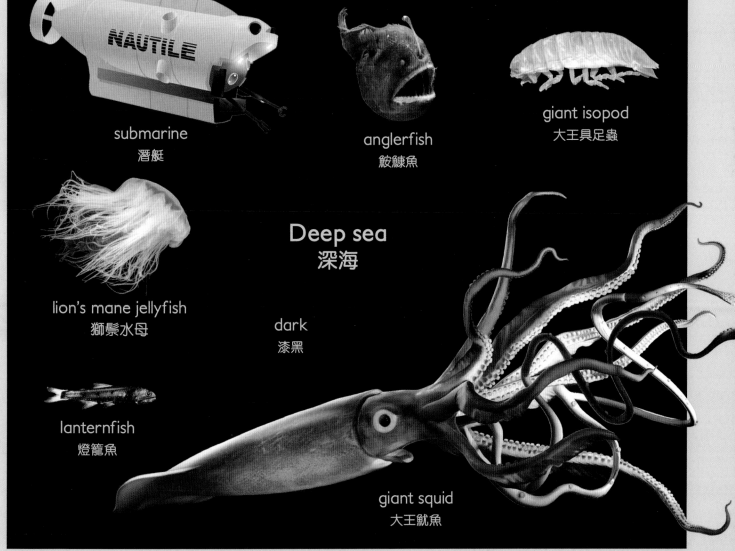

# Deep sea
# 深海

lion's mane jellyfish
獅鬃水母

dark
漆黑

lanternfish
燈籠魚

giant squid
大王魷魚

# Rainforests 雨林

Many animal species live in rainforests, and there are lots of new species still to be discovered. Let's look at what lives in the Amazon rainforest of South America.

在雨林裏，有許多不同種的動物，而且還有更多的新物種有待我們來發現。現在我們來看看生活在南美洲的亞馬遜雨林裏的生物吧。

high
高處

capuchin monkey
捲尾猴

sunny
陽光普照

kapok tree
木棉樹

climb
攀爬

sloth
樹懶

hang
懸掛

howler monkey
吼猴

vampire bat
吸血蝙蝠

## Emergent layer 露生層

fly
飛行

macaw
金剛鸚鵡

harpy eagle
角鵰

glide
滑翔

blue morpho butterfly
大藍閃蝶

tree boa
樹蚺

toucan
巨嘴鳥

green iguana
綠鬣蜥

What sounds might you hear in a rainforest?
在雨林裏，你可能聽見什麼聲音？

# Canopy
# 冠層

damp
濕潤

red-eyed
tree frog
紅眼樹蛙

bananas
香蕉

nutrient
rich
營養豐富

fungi
真菌

decay
腐爛

cocoa tree
可可樹

buttress roots
板根

cocoa pod
可可莢果

jaguar
美洲豹

# Understorey
# 下木層

giant
centipede
巨人蜈蚣

churo snail
蘋果螺

scorpion
蠍子

jewel beetle
吉丁蟲

orchid bee
蘭花蜂

leaf cutter
ant
切葉蟻

# Forest floor 地面表層

orchid
蘭花

armadillo
犰狳

giant
anteater
大食蟻獸

harlequin
beetle
長臂天牛

51

# Savannahs 稀樹草原

Savannahs are flat grasslands with few trees.
There are often wildfires, but plants regrow.
Some of the biggest savannahs are in Africa,
South America, and Australia.

稀樹草原是平坦的草原，生長的樹木很少。
稀樹草原常常發生山火，不過植物會重新生
長。部分面積最大的稀樹草原位於非洲、南
美洲及澳洲。

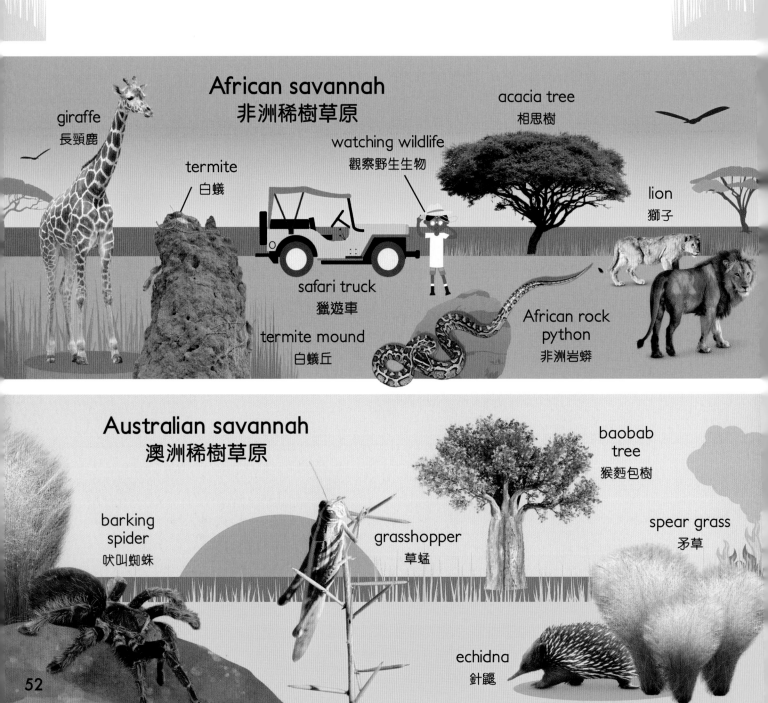

**African savannah**
**非洲稀樹草原**

giraffe
長頸鹿

termite
白蟻

watching wildlife
觀察野生生物

acacia tree
相思樹

lion
獅子

safari truck
獵遊車

termite mound
白蟻丘

African rock
python
非洲岩蟒

**Australian savannah**
**澳洲稀樹草原**

barking
spider
吠叫蜘蛛

grasshopper
草蜢

baobab
tree
猴麵包樹

spear grass
矛草

echidna
針鼴

# South American savannah
# 南美洲稀樹草原

pampas grass
蒲葦

rhea
美洲鴕

guinea pig
天竺鼠

pampas fox
河狐

puma
美洲獅

footprints
足印

Which big mammal has a very long nose?
哪一種大型哺乳類動物擁有非常長的鼻子？

elephant dung
大象糞便

impala
飛羚

zebras
斑馬

aardvark
土豚

watering hole
水坑

African elephant
非洲象

wildfire
山火

screwpine
露兜樹

eucalyptus
桉樹

kangaroo
袋鼠

dung beetle
糞金龜

frilled lizard
傘蜥蜴

possum
負鼠

bandicoot
袋狸

# Deserts 沙漠

Deserts can be hot or cold, but they all get very little rain. Let's look around the Chihuahuan Desert in Mexico.

沙漠可以是炎熱或寒冷的，不過它們全都只有非常稀少的雨水。一起來墨西哥的奇瓦瓦沙漠到處看看吧。

red-tailed hawk
紅尾鵟

hunt
捕獵

## Food chain 食物鏈

hiss
嘶嘶聲

rattle
咔察咔察聲

flow of nutrients
營養轉移

rattlesnake
響尾蛇

prickly pear cactus
刺梨仙人掌

jackrabbit
長耳大野兔

What eats what in the food chain?
在食物鏈中的生物吃什麼東西？

tumbleweed
風滾草

lizard
蜥蜴

coyote
郊狼

desert kit fox
敏狐

jaguar
美洲豹

dry
乾旱

barrel cactus
金琥

sand dune
沙丘

hot
炎熱

sun
太陽

arid
乾燥

golden eagle
金鵰

grasshopper
草蜢

dust storm
沙塵暴

scorpion
蠍子

bighorn sheep
大角羊

sand
沙

saguaro cactus
巨人柱仙人掌

Find three birds and two reptiles.
找出三隻鳥類和兩隻爬蟲類動物吧。

roadrunner
走鵑

bobcat
山貓

Joshua
tree
寬葉絲蘭

red-spotted toad
紅點蟾蜍

yucca
絲蘭

aloe
蘆薈

shade
陰影

desert ants
沙漠蟻

tarantula
狼蛛

55

# Mountains 高山

Mountains are towering lands of earth and stone. It gets colder as you go higher, so mountains are often snowy at the top.

高山是高聳的土地，由泥土與石塊組成；越高的地方氣溫越冷，因此高山頂端往往布滿冰雪。

golden eagle
金鵰

snow
雪

cold
寒冷

sky
天空

mountain goat
北美山羊

pine marten
松貂

hiking
登山

stone
石頭

foothill
山麓

lynx
猞猁

hiking boots
登山靴

wolf
狼

red squirrel
紅松鼠

reindeer
馴鹿

cave
洞穴

bear cub
幼熊

Where could the mother bear make her den?
母熊會在哪裏建造巢穴？

brown bear
棕熊

cloud
雲

summit
山頂

snowflakes
雪花

peregrine
falcon
遊隼

ski poles
滑雪杖

ski
滑雪

frozen
結冰

mountain
climber
登山者

skier
滑雪者

rope
登山繩

mountain
高山

tent
帳幕

camping
露營

cyclists
單車手

bicycle
單車

path
小徑

monarch
butterflies
帝王斑蝶

fish
魚

waterfall
瀑布

marmot
土撥鼠

plunge pool
瀑布池

rocks
岩石

# Arctic and Antarctic
## 北極與南極

Some amazing animals and plants have adapted to survive in the freezing polar habitats of the Arctic and the Antarctic.

有些動物和植物很奇妙，竟然能夠在北極和南極這些冰封極地中棲息存活。

The wildlife is different in each place. Do penguins live in the Arctic or the Antarctic?

在不同地方有不同的野生動物，你知道企鵝是居住在北極、還是南極嗎？

Arctic
北極

Arctic cottongrass
北極棉草

bog bilberry
黑豆樹

grayleaf willow
灰藍柳

reindeer lichen
石蕊

snowy owl
雪鴞

walrus
海象

Arctic tern
北極燕鷗

rocky shoreline
岩質海岸線

Arctic fox
北極狐

lemming
旅鼠

reindeer
馴鹿

Arctic hare
北極兔

polar bear
北極熊

beluga whale
白鯨

harp seal
豎琴海豹

ice
冰

frozen
結冰

narwhal
獨角鯨

# Antarctic
# 南極

polar lights
極光

Antarctic hair grass
南極髮草

Antarctic pearlwort
南極漆姑草

peak
山頂

mountain
高山

snow
雪

valley
山谷

wandering albatross
漂泊信天翁

snow petrel
雪鸌

iceberg
冰山

polar desert
極漠

orca
虎鯨(殺人鯨)

adelie penguin
阿德利企鵝

krill
磷蝦

leopard seal
豹海豹

Weddell seal
威德爾海豹

Spot three Antarctic mammals.
試找出三種南極的哺乳類動物。

gentoo penguin
巴布亞企鵝

emperor penguin
皇帝企鵝

starfish
海星

# Protecting nature 保護大自然

We should take care of our world. The choices we make can harm or protect nature.

我們要好好地保護地球，因為我們所作的抉擇，有可能對大自然做成傷害，又或可保護大自然。

Which of the things on these pages help to protect nature?
在這兩頁中，有哪些事情是有助保護大自然的？

## Our animals 我們的動物

conservation
保護

protected wildlife reserve
受保護野生生物保育區

anti-poaching unit 反偷獵小組

endangered
瀕危

reintroduction
再引入

caring
照顧

seagull
海鷗

extinct
滅絕

western black rhinoceros
西部黑犀

kites
鳶

zoo breeding programme 動物園繁殖計劃

## Our farming
## 我們的農業

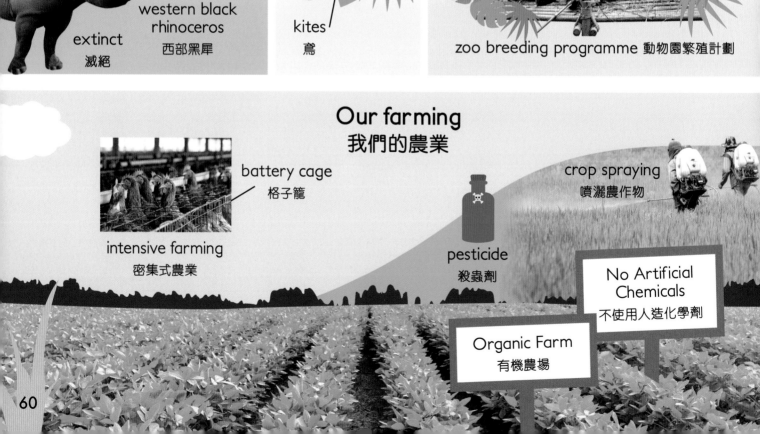

battery cage
格子籠

crop spraying
噴灑農作物

intensive farming
密集式農業

pesticide
殺蟲劑

No Artificial Chemicals
不使用人造化學劑

Organic Farm
有機農場

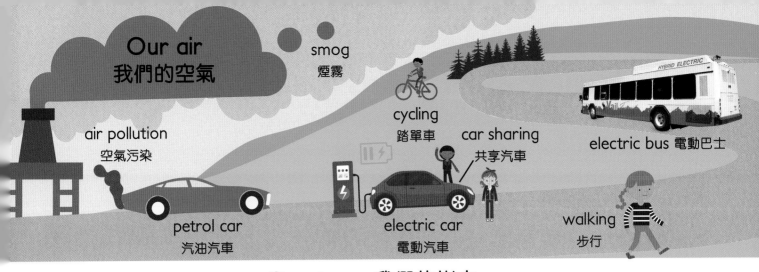

Our air
我們的空氣

smog
煙霧

air pollution
空氣污染

petrol car
汽油汽車

cycling
踏單車

car sharing
共享汽車

electric car
電動汽車

electric bus 電動巴士

walking
步行

## Our trees 我們的樹木

deforestation 砍伐森林

reforestation 再造林

## Our climate 我們的氣候

global
warming
全球暖化

Reducing 減少使用
Reusing 重用
Recycling 循環再造

wildfires
山火

floods
洪水

polar ice caps melting
極地冰蓋融化

Can we do something to help
nature today?
我們現今能做些什麼來保育大自然？

## Our oceans 我們的海洋

ocean litter
海洋垃圾

cleaning up
清潔海灘行動

# Working with nature

## 保育大自然

There are many ways to work with nature, including jobs in which you can help to protect our planet and the species that live on it.

保育大自然有許多方式，從事以下職業的人，都在保護地球和保育地球上的生物物種。

air quality scientist:
checks air pollution
空氣質素科學家：檢測空氣污染

farmer
農夫

geologist:
studies Earth's rocks
地質學家：研究地球的岩石

forester:
looks after forests
林務員：照顧森林

zoologist: studies animals
動物學家：研究動物

entomologist:
studies insects
昆蟲學家：研究昆蟲

environmental scientist:
studies the environment
環境科學家：研究環境

botanist: studies plants
植物學家：研究植物

gardener
園丁

tree surgeon
樹藝師

zookeeper
動物園管理員

marine biologist:
studies ocean life
海洋生物學家：研究海洋生物

vet
獸醫

environmental activist:
supports protecting
the environment
環境運動家：支持保護環境

natural history
museum
event manager
自然歷史博物館項目經理

ecologist: studies living
things and their environment
生態學家：研究生物和
它們的居住環境

street cleaner
街道清潔工

wildlife author
野生生物作家

wildlife photographer
野生生物攝影師

aquarist: manages
an aquarium
水族館館長：管理水族館

countryside ranger:
looks after the countryside
郊外護林員：照顧郊外

safari guide
獵遊導遊

wildlife presenter
野生生物導賞員

environmental engineer:
designs things for the
environment
環境工程師：為環境設計不同事物

animal rescue worker
動物拯救員

seismologist:
studies earthquakes
地震學家：研究地震

Which of these jobs
would you like to do?

在這些職業中，
你最想做哪一種呢？

63

# Acknowledgements

# 鳴謝

**DK would like to thank: Victoria Palastanga and Eleanor Bates for additional design work; Jagtar Singh for additional DTP design work; Adhithi Priya, Sakshi Saluja, Rituraj Singh, Sumedha Chopra, and Vagisha Pushp for additional picture research work; Polly Goodman for proofreading.**

**The publisher would like to thank the following for their kind permission to reproduce their photographs:**

**(Key: a=above; b=below/bottom; c=centre; f=far; l=left; r=right; t=top)**

1 123RF.com: Liubov Shirokova (clb). Dreamstime.com: Marc Bruxelle (cra); Isselee (tc); Vvoevale (tl); Yinan Zhang (cl); Svetlana Larina / Blair_witch (fclb); Nejron (cb); Sabelskaya (br). 2 Dorling Kindersley: Tom Grey (cla); Natural History Museum, London (tr). Getty Images / iStock: photo5963 (clb). 3 123RF.com: isselee / Eric Isselee (bc); Prapan Ngawkeaw (cb/road); jackf / Iakov Filimonov (bl/wolf). Dreamstime.com: Iakov Filimonov (clb/goat); Phartisan (rock x3); Jpsdk / Jens Stolt (butterflies x 3); Yotrak (cb/tent); Jackf / Iakov Filimonov (bc/reindeer); Nelikz (bl/squirrel). Shutterstock.com: Aleksandr Pobedimskiy (clb/sandstone). 6 – 7 Dreamstime.com: Gritsalak Karalak (c). 6 Dreamstime.com: Astrofireball (c). 7 Dorling Kindersley: Dan Crisp (ca); Natural History Museum, London (crb); James Kuether (cb); Jon Hughes (bc). Dreamstime.com: Nicolas Fernandez (c); Markus Gann / Magann (tr).

8 123RF.com: algre (crb). Dreamstime.com: Costasz (bc); Tomasz Smigla (bc/cups); Vchalup (br); Rui Matos / Rolmat (cra); Rob Wilson / Robwilson39 (cr). 9 123RF.com: julynx (tl); Martin Spurny (cla); Natthawut Panyosaeng / aopsan (ca). Alamy Stock Photo: Mouse in the House (bl). Dreamstime.com: Sergey Dzyuba (cr); Elena Kazanskaya (cla/bin); Tele52 (cra); Radha Karuppannan / Radhuvenki (clb/x2); Stockphototrends (bc). Getty Images / iStock: DigitalVision Vectors / bubaone (cl); photo5963 (c); youngID (ftl); pterwort (br). 10 123RF.com: Inna Astakhova (crb); tempusfugit (bc); nerthuz (cb). Dorling Kindersley: Tracy Morgan (cb/dog). Dreamstime.com: Melanie Hobson (ca/landscape); Nexus7 (t); Isselee (cl); Theo Malings (clb); Eric Isselee (cr). 11 123RF.com: Aaron Amat (clb/x3); Dmitry Rukhlenko / dimol (fcra). Alamy Stock Photo: Justin Kase z12z (fbl). Dreamstime.com: Billy Ber (tr); Narathip Ruksa / Narathip12 (ftr); Cammeraydave (cb); Cynoclub (bc); Jeroen Van Den Broek / Vandenbroek29 (bc/GuideDog); PhotoChur (br); Zbynek Burival / Merial (fbr). Fotolia: Eric Isselee (bl); Norman Pogson (ca).

Getty Images / iStock: Mac99 (cra); pifate (crb). 12 123RF.com: fireflamenco (x3). Dreamstime.com: Catalin205 (clb, crb); Vectorikart (cra); Stockoxinoxi (ca); Thomas Holt (cl); Macrovector (clb/net); Photka (br). 13 Dreamstime.com: Akinshin (tr); Vectorikart (cla); Andreanita (tr/bear); Miramisska (cra); Pavel Rodimov (c/stargazing); Kotenko (c); Gerald Zaffuts (c/storytelling); Christinlola (cr); Sabelskaya (clb); Kellyrichardsonfl (bl); Pavel Naumov (cb/x4); Sergiy Bykhunenko (br); Macrovector (ca). 14 Dreamstime.com: Andreiuc88 (cb); Antares614 (c); Nehru (br). 15 123RF.com: Rune Kristoffersen (cla). Dreamstime.com: Mihai Andritoiu (cb); Mishoo (tl); Artisticco Llc (tr).

16 123RF.com: justoomm (bc); Kitsadakron Pongha (cra); nasaimages (c/cyclone). Alamy Stock Photo: Mike Hill (c). Dreamstime.com: Arevhamb (bl); Andrey Armyagov (cla); Hulv850627 (cl); Justin Hobson (cr); Trekandshoot (clb); Thescv (cb); Elantsev (crb); Ruthchoi (br). 17 123RF.com: alicenerr (c). Alamy Stock Photo: SPUTNIK (cra). Dreamstime.com: Giuseppe Di Paolo (ca); Siempreverde22 (cla); Dmitry Pichugin / Dmitryp (ca/Everest); Valore (cl); Unissunil (c/Mawsynram); Jon Chica Parada (cr). Getty Images: Daniel Osterkamp (bc). Getty Images / iStock: Jorge Villalba (bl). 18 123RF.com: Ruth Jenkinson / stevanzz (bl); Andrzej Tokarski / ajt (tc); Przemyslaw Koch (fclb).

Dorling Kindersley: Liberty’s Owl, Raptor and Reptile Centre, Hampshire, UK (crb); Thomas Palmer (cb). Dreamstime.com: Digitalimagined (clb/liverwort); Anna Sedneva / Sedneva (clb/grass); Vitalssss (c); Sarah2 (crb/tick); Ildar Galeev (bc). Fotolia: Karl Bolf (fcrb). Getty Images / iStock: Antagain (clb). 19 123RF.com: smileus (tc); Pavlo Vakhrushev / vapi (cla); Ten Theeralerttham / rawangtak (cl); Thawat Tanhai (bl). Dreamstime.com: Conchasdiver (clb); Lgor Dolgov / Id1974 (tr); Ronniechua (c); Kotomiti_okuma (cb); Kazoka (cr). 20 Dreamstime.com: Cherdchai Chaivimol (cb/bud); Vaclav Volrab (cla); Kaiwut Niponkaew (ca); Tomboy2290 (cra/Basil); Natali572 (cra); Ppy2010ha (clb); Dewins (bc); Lepas (bl); Oleg Dudko (fbl); Bogdan Lazar (cb); Songyuth Unkong (crb); Mikhail Dudarev (br). 21 123RF.com: Anna Liebiedieva / utima (cl); olegdudko (c/Kiwi). Dorling Kindersley: Neil Fletcher (cra). Dreamstime.com: Anton Ignatenco (c); Zerbor (cla); David Ridley (ftl); Paul Rookes (tl); Zerbor (fcla); Natika (clb); Elena Schweitzer / Egal (cb); Roman Ivaschenko (bl); Roman Ivaschenko (bl/seaweed); Vetre Antanaviciute-meskauskiene (bc). Getty Images / iStock: DNY59 (fcra). Shutterstock.com: Daydreamr Digital Studio (tr). 22 Dorling Kindersley: Centre for Wildlife Gardening / London Wildlife Trust (ca). Dreamstime.com: Marc Bruxelle (cl); Vvoevale (cla); Pipa100 (clb); Dreamstock (cb); Filmfoto (bl); Anatoliy Mandrichenko (bc). 22 – 23 Dreamstime.com: Andreykuzmin (b); Zerbor (c). 23 Dreamstime.com: Denira777 (cra); Anton Ignatenco (tr); Ievgenii Tryfonov (bc); Setory (br); Majormetts (crb); Zorica Vitanovic (ca). 24 Dreamstime.com: Domnitsky (ca); Md. Rakibul Hassan (sunflower life cycle); Ilonai (tr). 25 Dreamstime.com: Elena Elisseeva (cr); Nadiia Havryliuk Kharzhevska (bl/mushroom cycle); Luayana (tl/apple tree life cycle); Angelo Gilardelli (tr); Lenazajchikova (crb); Pavel Rodimov (br); Wirestock (clb); Thawats (cla). 26 123RF.com: Aleksandr Ermolaev (cr). Dreamstime.com: Anankkml (bc); Photodeti (ca); Wirestock (br). Fotolia: Mark Higgins (fbr). Getty Images / iStock: LUNAMARINA (bl). 27 Alamy Stock Photo: SConcepts (bl). Dreamstime.com: Jason W. Baker (tl); Isselee (cra); Svetlana Larina / Blair_witch (ftl); Pimmimemom (tc); Stevenrussellsmithphotos (ca, fcla); Vasiliy Vishnevskiy (cl); Isselee (cr); Cinnamongirl (crb). Science Photo Library: Claude Nuridsany & Marie Perennou (tr). Shutterstock.com: Lamnoi Manas (cb). 28 123RF.com: Anna Utekhina (bc). Dreamstime.com: Accept001 (c); Isselee (bc/dalmatian); Judith Dzierzawa (br); Alexander Potapov (clb); Isselee (clb/sheep); Alexander Potapov (cra); Isselee (cra/cow); Tristana / Kseniya Abramova (ca); Isselee (cla). Fotolia: Anatolii (cb); Olena Pantiukh (crb). Getty Images: mikroman6 (bl). 29 123RF.com: Cathy Keifer (tc); smileus (tr). Alamy Stock Photo: Lee Dalton (clb). Dorling Kindersley: British Wildlife Centre, Surrey, UK (cla). Dreamstime.com: Linda Caldwell (bl); Toby Gibson (bc); Dizzizzmee (ca); Paul Farnfield (fcla); Steve Oehlenschlager (tl); Brett Hondow (tc/wolf spider). Fotolia: Eric Isselee (c). Getty Images: Photodisc / Don Farrall (ftl).

30 123RF.com: Visarute Angkatavanich (br). Dreamstime.com: Adogslifephoto (cla); Sutisa Kangvansap (bc); Prin Pattawaro (bl); Graeme Snow (crb); Billybruce2000 (cb); Marcin Wojciechowski (cr); Joanna Zopoth Lipiejko (c); Sonsedskaya (cl); Isselee (cra); Gianluca Piccin (ca). Getty Images / iStock: RAUSINPHOTO (clb). 31 Dreamstime.com: Mira Agron (fcl); Sebastian Kaulitzki (br); Pavel Trankov (bc); Steve Allen (bl); Isselee (crb); Isselee (cb); Isselee (clb); Gualberto Becerra (cr); Marco Tomasini (cl); Duncan Noakes (fcra); Wrangel (cra); Melinda Fawver (ca); Svetlana Foote (cla); Dirk Ercken (fcla); Jacoba Susanna Maria Swanepoel (ftr); Wildlife World (tr); Jagodka (tc); Alexandercreator (tl); Reimarg (ftl). 32 123RF.com: ksena32 / Oksana Tkachuk (cra/chamomile). Alamy Stock Photo: Jason Bazzano (br); mauritius images GmbH / Kurt Madersbacher (crb). Dreamstime.com: Andreanita (cra); Jose Manuel Gelpi Diaz (cra/vulture); Valentyna Chukhlyebova (ca, cr); Hwongcc (cla, c); Golfxx (cb); Marazem (bc); Dmitry Potashkin (bl); Pop Nukoonrat (sky). 33 Dreamstime.com: Kharis Agustiar (cr); Jocrebbin (bc); Zedcreations / SACHITH (webs x2); Geerati (cra); Duncan Noakes (cla). Getty Images: The Image Bank / Joe McDonald (bl). Getty Images / iStock: Parrotstarr (clb); superjoseph (ca); pixhook (ca/bamboo). 34 123RF.com: Isselee Eric Philippe (c). Alamy Stock Photo: Arterra Picture Library / Clement Philippe (br). Dreamstime.com: Callipso88 (cr); Vasyl Helevachuk (clb); Dizm (cb); Isselee (crb); Wildlife World (ca); Corey A Ford (cla). Getty Images: Photodisc / Don Farrall (bc). Shutterstock.com: A.Mac.Photo (cl). 35 123RF.com: John McAllister (cl); utima (ca). Alamy Stock Photo: Bill Coster (c); Pally (br); Les Gibbon (bl). Dreamstime.com: Angel Luis Simon Martin (crb); Sergeyoch (clb). Getty Images: Hearst Newspapers / San Francisco Chronicle (cra). Getty Images / iStock: LuckyTD (tc). 36 123RF.com: swavo (br); Andrzej Tokarski / ajt (c). Dorling Kindersley: Natural History Museum, London (cr). Dreamstime.com: Isselee (clb); Alexander Konoplyov (br/bacteria); Stu Porter (cra); Mrrphotography (tl). Getty Images / iStock: bbevren (b). 37 123RF.com: Eric Isselee (crb). Ardea: ar / Science Source / Tom McHugh (cl).

Dorling Kindersley: Natural History Museum, London (c); Jerry Young (bl); Wildlife Heritage Foundation, Kent, UK (tl). Dreamstime.com: Isselee (clb); Jblackstock / Justin Black (cb); Yves Sautter (c/octopus). 38 123RF.com: Eric Isselee (crb). Dreamstime.com: Nchuprin / Andrey Sukhachev (crb/bacteria); Angelique Nijssen (bl); Thatsaphon Saengnarongrat (tr). 38 – 39 Dreamstime.com: Andreykuzmin (soil); Smishko (sand texture). 39 123RF.com: Sayompu Chamnankit (bc/footprints). Alamy Stock Photo: Rosanne Tackaberry (br). Dorling Kindersley. Dreamstime.com: Andreykuzmin (b); Dijarm (br/graph); Kosmos111 (crb); С е р г е й К у ч у г у р н ы й (bl); Tamara Kulikova (cb); Hommalai (tc); Dave Nelson (ca); Thanthip Homsansri (cla); Lcrms7 (c); Sripfoto (cl); Typsiaod (clb). Shutterstock.com: Mikkola (cr). 40 123RF.com: alekss / Alexandr Pakhnyushchyy (bl); Anatolii Tsekhmister (cra). Alamy Stock Photo: Don Despain (clb). Dorling Kindersley. Dreamstime.com: Chernetskaya (fcrb); Uros Petrovic (br); Pzaxe (bc); Loren File (crb); Ivonne Wierink (fcrb/pots); Jgade (cla). Shutterstock.com: Artiste2d3d (fcra).

41 Dreamstime.com: Nikolay Antonov (clb/worm); Kristof Lauwers (br); Pimmimemom (crb); Aleksandr Volkov (clb); Fibobjects (cb); Luceluceluce (clb/soil); Atlasfotoreception (crb/gloves); Boulanger Sandrine (cr); Bundit Minramun (cla); Sergiy1975 (cla/lawn mower). 42 – 43 Dreamstime.com: Miriam Doerr (flower x3); Eugenesergeev (grass); Miriam Doerr (wild flowers x3); Supertrooper (c). 42 123RF.com: peterwaters (clb/bee). Alamy Stock Photo: Life on white (cla/horse). Dorling Kindersley: Mark Hamblin (cra). Dreamstime.com: Animaflora (cb); Tazzymoto (ca); Brett Critchley (cla); Isselee (c); Dzmitry Shpak (fcrb); Inna Kyselova (crb); Nipaporn Panyacharoen (cb/barley); Thawats (bc). Shutterstock.com: Volosina (clb). 43 Alamy Stock Photo: Islandstock (cla). Dorling Kindersley: Twan Leenders (cb/snake). Dreamstime.com: Tony Bosse (tr); Mickem (bc); Tchara (cb); Chuyu (clb); Sandra Standbridge (crb); Isselee (c); Romica (cra); Palians (cra/harvester); Mariya Kondratyeva (cra/land). 44 123RF.com: Eric Isselee (c). Alamy Stock Photo: Imagebroker / Arco / J. Fieber (bl). Dorling Kindersley: Roger Tidman (cl). Dreamstime.com: ActiveLines; Macrovector (clb); Isselee (crb); Stephanie Frey (cr); Zerbor (r); Atman (tc); Mille19 (clb/owl). Getty Images / iStock: MarkMirror (cb). 45 123RF.com: Eric Isselee (cra/koala). Dreamstime.com: Karen Black (cra); Alexander Potapov (bc); Susan Sheldon (bl); Geoffrey Kuchera (bl/bear); Donyanedomam (clb); Lunamarina (cla); David Steele (ca). Getty Images / iStock: GlobalP (cb). Shutterstock.com: aphotostory (crb). 46 – 47 Dreamstime.com: David Watson (bc). Shutterstock.com: xpixel (cane x5). 46 123RF.com: Stanko Mravljak (crb/mayfly). Dorling Kindersley. Dreamstime.com: G3miller / Gordon Miller (cr); Zerbor (tr); Eduard Kyslynskyy (ca); Dennis Jacobsen (c); Kevin Wells (cb); Photophreak (crb); Roman Ivaschenko (bl). Shutterstock.com: Igor Podgorny (cla). 47 123RF.com: Stefan Holm (clb/dragonfly); NewAge (cl). Dorling Kindersley: Roger Tidman (c). Dreamstime.com: Natalya Aksenova (ca); Sova004 (br); Ilyas Kalimullin (crb); Kobchaima (cra); Wirestock (tr); Jnjhuz (clb); Zeytun Images (cb). 48 – 49 Getty Images / iStock: photo5963 (ca). 48 Alamy Stock Photo: Doug Perrine (cr); SBS Eclectic Images (cb); Carsten Reisinger (crb); WaterFrame_dpr (clb). Dorling Kindersley: Natural History Museum, London (bl); Linda Pitkin (cl). Dreamstime.com: Kevin Panizza (fclb); Pipa100 (cb/lettuce); Harvey Stowe (cra). Fotolia: uwimages (crb/anemonefish). 49 123RF.com: feathercollector (cr). Alamy Stock Photo: Minden Pictures / Norbert Wu (c). Dorling Kindersley: Tom Grey (tc). Dreamstime.com: Robertlasalle (clb). naturepl.com: Solvin Zankl (clb/lanternfish). 50 – 51 Dreamstime.com: Surachet Khamsuk. 50 123RF.com: Hal Brindley (tc); Hal Brindley (fcra). Alamy Stock Photo: Ivan Kuzmin (tr); Nature Picture Library / MYN / Andrew Snyder (crb); Nature Picture Library / Nick Garbutt (cb/eagle). Dreamstime.com: Carlosphotos (bc); Nejron (cb); Chansom Pantip (ftr); Arindam Ghosh (cra). naturepl.com: Luiz Claudio Marigo (cla). 51 123RF.com: anankkml / Anan Kaewkhammul (cb/jaguar); Ajay Bhaskar (r). Alamy Stock Photo: Zizza Gordon Insect collection (bl). Dreamstime.com: Beautifulblossom (tr); Ryszard Laskowski (bc); Gan Chaonan (br); Whiskybottle (bc/orchid); Isselee (crb); Olga Soe (red flowers x3); Thenatureguy1 (clb); Morley Read (crb/scorpion); Vlad Ivantcov (c); Superoke (cla); Douglas Delgado (cra); Ekays (tl); Waraphot Wapakphet (tl/leaves). naturepl.com: Gabriel Rojo (cb). 52 123RF.com: waldemarus (bc/baobab). Dreamstime.com: Bennymarty (crb); Yinan Zhang (cl); Snehitdesign (c); Alexander Shalamov (cr); Kewuwu (c/tree); Fritz Hiersche (bc); Svetlana485 (br); Alexandr Yurtchenko (tc). naturepl.com: Piotr Naskrecki (clb/ant 1, clb/ant 2, clb/ant 3). 53 Alamy Stock Photo: Ken Griffiths (br). Dorling Kindersley: Blackpool Zoo (clb); Wildlife Heritage Foundation, Kent, UK (cra). Dreamstime.com: Anekoho (cr); Lev Kropotov (crb); Pokec / Jan PokornÃ (cb); Birdiegal717 (bc); Izanbar (bc/possum); Godruma (bl); Johan63 / Johannes Gerhardus Swanepoel (cl); Luca Santilli (c); Vicspacewalker (cla); Luciano Queiroz (ca); Rafael Cerqueira (cla/guinea pig). 54 123RF.com: cookelma / Andrey Armyagov (b); Anan Kaewkhammul / anankkml (br). Alamy Stock Photo: George Brice (bl); Robert Shantz (cb); Mike Lane (cra). Dorling Kindersley: Andy and Gill Swash (cr). Dreamstime.com: Steve Byland (tl); Derrick Neill (clb); Eutoch (fcr); Eutoch (fcra); Isselee (cla). 55 123RF.com: alhovik (fcra); Natalie Ruffing (cr); sladerer / Scott Laderer (ca); Ufuk Zivana (ca/cactus). Alamy Stock Photo: Arterra Picture Library / Clement Philippe (bc); Nature Picture Library / John Abbott (bl); Rick & Nora Bowers (crb). Dreamstime.com: Dfikar (clb); Frank Fichtmueller (fcrb); Skynetphoto (cra); Domnitsky (c); Withgod / Alexander Podshivalov (tl); Vally (cla). 56 – 56 Dreamstime.com: Phartisan (rock x4). 56 – 57 Shutterstock.com: Aleksandr Pobedimskiy (br/sandstone x2). 56 123RF.com: jackf / Iakov Filimonov (cb). Alamy Stock Photo: Niebrugge Images (br); Paulette Sinclair (cb/bear); Ronald S Phillips (bc). Dreamstime.com: Jim Cumming (cra); Jackf / Iakov Filimonov (crb); Iakov Filimonov (ca); Nelikz (clb). 57 123RF.com: isselee / Eric Isselee (clb); Prapan Ngawkeaw (cl). Dreamstime.com: Jpsdk / Jens Stolt (butterflies x3); Yotrak (cla). Shutterstock.com: Yes058 Montree Nanta (br/granite). 58 Alamy Stock Photo: All Canada Photos / Wayne Lynch (fclb); Dembinsky Photo Associates / Alamy / Dominique Braud (cb); Bob Gibbons (cra); Realimage (cla). Dorling Kindersley: Jerry Young (crb). Dreamstime.com: Devon Crosby (bl); Planetfelicity (br); Luna Vandoorne Vallejo (bc); Grafner (b); Outdoorsman (fcrb); Uhg1234 (clb/reindeer); Zanskar / Vladimir Melnik (cla/walrus); Luis Leamus (cr); Il'mar Idiyatullin (fcra); Troyka (ca). Getty Images / iStock: Karyn Schiller (fcla). 59 Alamy Stock Photo: era-images / Colin Harris (tr); Minden Pictures / Norbert Wu (bc). Dreamstime.com: Agami Photo Agency (cla); Photographerlondon (br); Sharon Jones (crb); Vika Ivanets (cb); Freezingpictures / Jan Martin Will (cr); Slowmotiongli (c); Staphy (ca); Biletskiy (t); Viktoria Ivanets (tc). Getty Images: Digital Vision / David Tipling (bl). 60 Alamy Stock Photo: Roger Hutchings (cra); WhiskeyWolf (cl); Ann and Steve Toon (c). Dreamstime.com: Adogslifephoto (ca); Chuchart Duangdaw (crb); Biletskiy (b); Comzeal (clb); Hel080808 (cr); Nilanjan Bhattacharya (cla/tiger); Maxirf (cra/anti poaching unit); Sarayut Thaneerat (cla). 61 Dreamstime.com: Steve Allen (bc); Skylightpictures (cb); Andrey Koturanov (cb/Waterflood); David Pereiras Villagra (br); Romolo Tavani (crb); Win Nondakowit (bl); Gpgroup (clb); Sjors737 (cra); Piotr Wawrzyniuk (cla); Noamfein (tr). 64 Dreamstime.com: Marc Bruxelle (tr/Maple); Vvoevale (tr); Nadiia Havryliuk Kharzhevska (crb); Zerbor (tr/maple tree)

Cover images: *Front and Back:* Dreamstime.com: Irinav; *Front:* 123RF.com: Aaron Amat bc/ (Ostrich), jackf / Iakov Filimonov cl, madllen (sprout), Liubov Shirokova (Flower), Andrzej Tokarski / ajt fbr, Anatolii Tsekhmister (squirrel); Dorling Kindersley: Blackpool Zoo cla, Centre for Wildlife Gardening / London Wildlife Trust (Hollyleaf), Mark Hamblin tc, Liberty’s Owl, Raptor and Reptile Centre, Hampshire, UK (tarantula), Natural History Museum, London (butterfly), Jerry Young (Bumblebee); Dreamstime.com: Atman (leaf), Marc Bruxelle (MapleLeaf), Carlosphotos (Butterflyx2), Denira777 cr, Lgor Dolgov / Id1974 cra, Dreamstock (Fir), Dvrcan cra/ (weevil), Iakov Filimonov cla/ (goat), Angelo Gilardelli br, Godruma ca/ (beetle), Vasyl Helevachuk (robin), Eric Issel é e (Silkworm), Isselee (Deer), Jblackstock / Justin Black tr, Jgade (frog), Johan63 / Johannes Gerhardus Swanepoel (impala), Svetlana Larina / Blair_witch clb, Nejron (parrot), Matee Nuserm fcrb, Pokec / Jan PokornÃ clb/ (Kangaroo), Stu Porter (Cheetah), Alexander Potapov (agaric), Stevenrussellsmithphotos tc/ (Butterfly), Ievgenii Tryfonov cl/ (trunk), Vasiliy Vishnevskiy cb/ (Rook), Vvoevale (brown leaf); Getty Images: Fuse cb/ (Jaguar); Getty Images / iStock: GlobalP tl, igorkov (eagle); *Back:* 123RF.com: madllen (sprout), Liubov Shirokova (Flower), Anatolii Tsekhmister (squirrel); Dorling Kindersley: Jerry Young (stickleback fish), Centre for Wildlife Gardening / London Wildlife Trust (Hollyleaf), Twan Leenders tl, Liberty’s Owl, Raptor and Reptile Centre, Hampshire, UK (tarantula), Natural History Museum, London (butterfly), Jerry Young (Bumblebee); Dreamstime.com: Atman (leaf), Marc Bruxelle (MapleLeaf), Dreamstock (Fir), Dvrcan (weevil), Freezingpictures / Jan Martin Will (penguin), Vasyl Helevachuk (robin), Eric Issel é e (Silkworm), Isselee (Deer), Jgade (frog), Johan63 / Johannes Gerhardus Swanepoel (impala), Svetlana Larina / Blair_witch ca, Luis Leamus cr, Nejron (parrot), Uros Petrovic crb, Stu Porter (Cheetah), Ievgenii Tryfonov (trunk), Vasiliy Vishnevskiy (Rook), Vvoevale (brown leaf), Zerbor bc; Getty Images / iStock: GlobalP (panda), igorkov (eagle); *Spine:* Dreamstime.com: Macrovector (snail), Alexander Potapov (agaric)

For further information see: www.dkimages.com